Vapour and Trace Detection of Explosives
for Anti-Terrorism Purposes

NATO Science Series

A Series presenting the results of scientific meetings supported under the NATO Science Programme.

The Series is published by IOS Press, Amsterdam, and Kluwer Academic Publishers in conjunction with the NATO Scientific Affairs Division

Sub-Series

I. **Life and Behavioural Sciences**	IOS Press
II. **Mathematics, Physics and Chemistry**	Kluwer Academic Publishers
III. **Computer and Systems Science**	IOS Press
IV. **Earth and Environmental Sciences**	Kluwer Academic Publishers
V. **Science and Technology Policy**	IOS Press

The NATO Science Series continues the series of books published formerly as the NATO ASI Series.

The NATO Science Programme offers support for collaboration in civil science between scientists of countries of the Euro-Atlantic Partnership Council. The types of scientific meeting generally supported are "Advanced Study Institutes" and "Advanced Research Workshops", although other types of meeting are supported from time to time. The NATO Science Series collects together the results of these meetings. The meetings are co-organized bij scientists from NATO countries and scientists from NATO's Partner countries – countries of the CIS and Central and Eastern Europe.

Advanced Study Institutes are high-level tutorial courses offering in-depth study of latest advances in a field.
Advanced Research Workshops are expert meetings aimed at critical assessment of a field, and identification of directions for future action.

As a consequence of the restructuring of the NATO Science Programme in 1999, the NATO Science Series has been re-organised and there are currently Five Sub-series as noted above. Please consult the following web sites for information on previous volumes published in the Series, as well as details of earlier Sub-series.

http://www.nato.int/science
http://www.wkap.nl
http://www.iospress.nl
http://www.wtv-books.de/nato-pco.htm

Series II: Mathematics, Physics and Chemistry – Vol. 167

Vapour and Trace Detection of Explosives for Anti-Terrorism Purposes

edited by

Michael Krausa

Fraunhofer Institut für Chemische Technologie,
Pfinztal (Berghausen), Germany

and

Aleksey Alekseyvitch Reznev

SMC Technological Center MIET,
Moscow, Russia

Kluwer Academic Publishers

Dordrecht / Boston / London

Published in cooperation with NATO Scientific Affairs Division

Proceedings of the NATO Advanced Research Workshop on
Vapour and Trace Detection of Explosives for Anti-Terrorism Purposes
Moscow, Russia
19–20 March 2003

A C.I.P. Catalogue record for this book is available from the Library of Congress.

ISBN 1-4020-2714-1 (HB)
ISBN 1-4020-2716-8 (e-book)

Published by Kluwer Academic Publishers,
P.O. Box 17, 3300 AA Dordrecht, The Netherlands.

Sold and distributed in North, Central and South America
by Kluwer Academic Publishers,
101 Philip Drive, Norwell, MA 02061, U.S.A.

In all other countries, sold and distributed
by Kluwer Academic Publishers,
P.O. Box 322, 3300 AH Dordrecht, The Netherlands.

Printed on acid-free paper

TABLE OF CONTENTS

Preface

The fast detection of explosives from the vapor phase would be one way to enhance the protection of society against terrorist attacks. Up to now the problem of detection of explosives, especially the location of explosives whether at large areas e.g. station halls, theaters or hidden in cars, aircraft cargo, baggage or explosives hidden in crowds e.g. suicide bombers or bombs in bags has not been solved. Smelling of explosives like dogs do seems to be a valuable tool for a security chain.

In general different strategies can be adopt to the basic problem of explosive detection:

- bulk detection

- vapor detection

Normally meetings cover both aspects and applications of the detection. Even though both methods might fulfill special aspects of a general security chain the underlying scientific questions differ strongly. Because of that the discussions of the scientists and practitioners from the different main directions are sometimes only less specific. Therefore the NATO Advisory Panel in Security-Related Civil Science and Technology proposed a small series of NATO ARW's which focuses on the different scientific aspects of explosives detection methods. This book is based on material presented at the first NATO ARW of this series in Moscow which covered the topic: Vapor and trace detection of explosives. The second ARW was held in St. Petersburg and treated the topic Bulk detection methods. The third workshop was held in Warwick and focused on electronic noses which cover a somewhat different aspect of vapor detection.

Up to now only sniffer dogs are able to search large areas and to locate explosives on these areas or hidden places by smelling the explosive. The main problem of vapor detection of explosives is the very small vapor pressures of the explosives. For example the vapor pressure of trinitrotoluene (TNT) is 7 ppb at room temperature. The problem becomes worse if we take into account that the real vapor concentrations of explosives are decreasing with increasing distance from the source. There are different strategies for development of sensor systems to solve this problem: enhancement of the sensitivity or sampling. Both aspects are discussed during the meeting and both ways might be helpful for different applications as part of a security chain.

Although sniffer dogs are able to locate explosives, RDX and HMX with smaller vapor pressures than TNT, but it is under discussion what dogs really smell. The answer to this question might give a very important input for the developer of sensors and/or electronic noses. During the workshop this question was intensively discussed and results were presented that dogs possibly did not smell the pure explosives but byproducts or odor bouquets of the explosive mixtures. There is a long tradition to train dogs as sniffer dogs for various applications and because of that a lot of information is available. Especially the scientific exchange between scientists with experience of sniffer dogs and sensor developer seemed to be very important and might speed up the development time of chemical sensors for vapor detection. One of the workshop goals was to bring scientists of both directions dogs handler and developer in close contact.

Beside the small concentrations of the explosives the variety of different and new energetic materials or home made explosives is an additional problem for chemical sensors systems. In view of commercial and military used explosives tagging and marking with substance compared to the pure explosive much higher vapor pressure is a good solution for the detection of the explosive. The point of tagging was discussed also as the fact how to react on home made explosives and new threats. In this sense the importance of the further communication between practitioners and sensor developer was emphasized.

During the final discussion it was stated that in view of counter-terrorism the vapor detection of explosives is a very important tool. At present chemical sensor systems are not able to fulfill all expected and wanted tasks but today systems might be sensitive and selective enough so that they could be integrated for special applications as part of a general security chain.

Terrorism nowadays is an international problem and therefore the communication between scientists is important to enhance the security. The intensive and open communication between scientists and experts from east and west with their knowledge and experience might help to find new ways and strategies in explosive detection for application in counter-terrorism.

Michael Krausa

Aleksey Alekseyvitch Rezev

Acknowledgement

The basis for a successful workshop and for exciting discussions is a good organization and a pleasant atmosphere of a workshop. During the time of the workshop I personally felt a very warm and friendly hospitality of the local organizers and the SMC Technological Center MIET in Moscow. This important basis was laid by the local organization team, namely: Dr. Anatoly Kovalev and Dr. Igor Verner. I was pleased that the director of the SMC Dr. Alexander Saourov and the rector Dr. Y. Chapligin gave us the possibility to hold this workshop at their institute. Moreover it was a great pleasure that Prof. Dr. Alexei Reznev, a member of the ICAO, was willing to take over the co-director tasks of the Russian side.

Another point for a successful workshop is a number of interesting lectures and the following discussions between the participants. I was very surprised at the intensity of the discussions, which I seldom attend at workshops. I felt that all participants wanted to continue the discussions but the time was appointed to two days only. I very hope that we will have more time during the next workshop. I like to thank all participants of the workshop for their contributions and their discussions contributions.

I was pleased that the NATO Advisory Panel on Security-Related Civil Science& Technology, especially the Programme Director Prof. F.C. Rodrigues gave me the chance for the organization of this workshop. In addition I would like to thank Prof. Hiltmar Schubert who gave me his support in preparing this workshop.

At conclusion please let me personally thank all participants, organizers and NATO panel members. For me it was a premiere to organize an ARW and you all made this performance a very pleasant and exciting experience for me. Thank you very much!

Michael Krausa

VAPOR DETECTION OF EXPLOSIVES FOR COUNTER-TERRORISM

MICHAEL KRAUSA

Fraunhofer-Institut für Chemische Technologie (ICT), Joseph-von-Fraunhofer-Str. 7, 76327 Pfinztal, Germany; Tel.: 0049-721-4640 444; e-mail: michael.krausa@ict.fhg.de

INTRODUCTION

The September 11th disaster on the twin towers was one of the most horrible terrorist attacks worldwide. In consequence of this attack a worldwide intensive discussion started concerning terrorism and the ways to protect society against terrorism. In the context of this debate the detection of explosives in an early stage of terrorist approach is one of the prominent chances for protection. But up to now several draw backs hamper the fast and easy detection of explosives and it is assumed that a 100 % guarantee to detect all explosives can never be reached. On the other hand an enhanced knowledge about the chemical behavior of explosives and byproducts in a real environment would give a vital input for research and development of more increasingly powerful sensor systems for the detection of explosives.

In principal two major approaches to detect explosives can be distinguished:
- bulk detection
- vapor detection

Today bulk detection of explosives is very common e.g. in airport security. Normally the luggage at airports is effectually controlled by X-ray systems. Several new methods are planned to enhance the selectivity and sensitivity of this proceeding. Otherwise humans could not be controlled in this way because of e.g. health risks and in addition x-ray systems usually are large and heavy systems and therefore it is difficult to transport them. Therefore these systems can't be used for the search of large areas (e.g. airport or station hall) or in the case of suicide bombers e.g. at theaters or bus stations. For these applications, search of large areas or personal control, so called sniffer dogs are the most successful explosive detection "systems" at present. Dogs are able to detect explosives from the vapor phase under various conditions. On the other hand dogs are living beings and their behavior and skills

1

M. Krausa and A. A. Reznev (eds.),
Vapour and Trace Detection of Explosives for Anti-Terrorism Purposes, 1-9.
© 2004 *Kluwer Academic Publishers. Printed in the Netherlands.*

are influenced by numerous parameters (e.g. age, sex, trainer, daily condition, etc.). To overcome these problems concerning the detection of explosives and illicit substances by dogs chemical sensors seem to be a valuable and important alternative for these applications. Moreover observations of the dogs strategies to detect explosives gives an important input for research and development of sensor systems for vapor detection.

Independent of the system, which are used for vapor detection, dogs included, some questions could be formulated which show a similar effect for all systems:

- Which explosives are used?
- How do environmental conditions influence the vapor phase?
- What does a dog smell?
- What are the consequences for sensor development or application?

WHICH EXPLOSIVES ARE USED?

Chemical sensors normally should be highly sensitive and very selective. In view of e.g. airport security selectivity is a very important condition because of the restriction of a low number of false alarms. On the other hand if the method is very selective only one chemical substance will be detected. Dogs might be trained to detect more than a single substance but the number of substances which could be detected by one dog is limited also. This means that for each substance which has to be detected a special chemical sensor has to be used. This is one of the most outstanding problems in view of explosives detection for counter-terrorism because it can not be predicted which explosive is used. Therefore information concerning explosives which are used in terrorist attacks or which are expected to be used for terrorist attacks are important for the development of new sensor systems.

One category of explosives which could be used and which are used in terrorist attacks are commercial available explosives. *Tab. 1* gives an overview about commercial explosives and their compositions. As can be seen from *Tab. 1* these explosives consist of a small number of different substances only [1]. Most of them could be detected e.g. from vapor phase samples by ion mobility spectroscopy (IMS) but at present not online from vapor phase. Furthermore it is conceivable that these chemicals could be measured by a combination of different chemical sensor systems which show a high sensitivity and selectivity for each specific substance respectively. In this case another problem arises which is connected with the small vapor pressures of these substances as will be presented in a later chapter.

But in general it has to be asked if these explosives *(Tab.1)* are normally used in terrorist attacks. The attack on the World Trade Center in 1993 was carried out by the use of nitrated urea, in Oklahoma City 1995 ammonium nitrate fuel oil (ANFO) was used, in Moscow 1999 presumable cyclonite was used and it is very difficult to predict what a suicide bomber will use. This problem becomes more complex if a statistic of Simmons is taken into account [2]. Simmons collected data about the chemicals which were used in different cases (criminal and terrorist). As can be seen from *Tab. 2* only in 3 % of all cases high explosives were used. 32 % of the bombings were conducted with smokeless or black powder, 29 % were conducted with simple chemical mixtures and 16 % were conducted by the use of pyrotechnic compositions.

Tab. 1: Commercial explosives and their main components [1]

Commonly used explosives	Main components
C-2	RDX+TNT+DNT+NC+MNT
C-3	RDX+TNT+DNT+Tetryl+NC
C-4	RDX+Polyisobutylene+Fuel oil
Cyclotol	RDX+TNT
DBX	TNT+RDX+AN+Al
HTA-3	HMX+TNT+Al
Pentolite	PETN+TNT
PTX-1	RDX+TNT+Tetryl
PTX-2	RDX+TNT+PETN
Tetroyl	TNT+Tetryl
Dynamite 3	NG+NC+SN
Red Diamond	NG+EGDN+SN+AN+Chalk

Tab. 2: Which explosives are used? (from R. Simmons, Naval Surface Warfare Center [2])

32 %	of the bombings used smokeless or black powder
29 %	used simple chemical mixtures (simple gas-producing chemicals confined in a container capable of withstanding some pressure before bursting)
16%	commercial fireworks or pyrotechnic composition similar to those used in fireworks
3%	were high explosives or ammonium nitrate blasting agents
14%	could not be identified

It is getting worse if home or self made explosives are additionally taken into account. Various broadcast stations shared the sensational news that documents were discovered near an Al Qaeda home in Kabul that contained hand-written notes how to make RDX and other explosives [3]. It is astonishing why this news was taken so sensational because the synthesis of numerous explosives and explosives mixtures are available in the World Wide Web [4,5]. So each person who has the ability to use the www has the chance to get the necessary information to synthesize explosives and to build bombs. Moreover there are numerous different instructions for several inorganic and organic explosives and new explosives and detonators e.g. triacetone triperoxide (TATP), which is a terrible threat, available in the www. Therefore the unmanageable number of different explosives makes it difficult to detect all of them and that is one of the greatest problems in view of explosives detection by common sensor systems (bulk and vapor) for counter terrorism.

On the other hand sniffer dogs are also able to find various home made explosives if the dogs are trained to detect the smell of these mixtures.

HOW DO ENVIRONMENTAL CONDITIONS INFLUENCE THE VAPOR PHASE?

Beside the unmanageable large number of different explosives several other factors complicate the explosive detection from the vapor phase. *Tab.3* shows the vapor pressures of some explosives. As can be seen the vapor pressures of these substances are comparatively small. That means sensor systems have to be very sensitive to detect these small amounts from the vapor phase. The problem becomes worse if it is taken into account that the vapor concentration of the explosives decrease strongly with increasing distance from the source. The vapor pressures of the substance therefore are the highest possible concentrations only. This problem becomes worse for inorganic materials, e.g. nitrated urea, potassium perchlorate, cane sugar etc. The vapor pressures of these chemicals are mostly much smaller than the pressures of the organic explosives. Therefore sensors have to be very sensitive if they focus on the detection of the pure explosive alone.

But the problem becomes more complicated by the influence of the environment and the physical behavior of different substances. For example the vapor pressure of TNT is small but beside of that TNT adsorbs strongly on nearly all materials and desorption is low. Information about the adsorption

Tab. 3: Vapor pressures of some explosives

Explosives	Vapor pressure (Torr) at 25°C
EGDN	2.8×10^{-2}
NG	4.4×10^{-4}
TNT	7.1×10^{-6}
PETN	1.4×10^{-8}
RDX	4.6×10^{-9}

and desorption of TNT is very rare especially for investigations under realistic conditions, e.g. TNT packaged in a plastic bag and packed in a suitcase which is full of clothes. How high will the TNT-concentration outside the suitcase be if any? On the other hand sniffer dogs are able to detect TNT under these conditions also. So the question arises what dogs really smell and additionally which kind of sensor do we need.

WHAT DOES A DOG SMELL?

It is still a controversial discussion what dogs smell. Beyond doubt the smell sense of dogs is stronger skilled than the sense of humans. Investigations which were conducted at the Canine Olfactory Detection Laboratory at the Auburn University showed that dogs are able to smell concentration as low as 10^{-12}-10^{-13} g [6]. More important in view of development of new sensors for the detection of explosives is the fact that dogs do not necessarily detect the pure explosive but byproducts or impurities. As an example at the Auburn University dogs were trained to identify nitro-glycerin based smokeless powder. The vapor phase of the powder consists of 80 different chemical substances. After the training the dogs smelled a pretended vapor phase which was composed of different substances and it was found that the dogs use an odor signature of acetone, toluene and limonene to identify the powder. They did not use the pure nitro-glycerin! [7]

In addition dogs are used today for mine detection in a very successful way. But the same questions arise in this case: What do the dogs smell? Or how smells a mine. At Auburn University the vapor phase above a PMA1-A landmine was investigated [8]. It is astonishing but no TNT was found but DNT, DNB, toluene, etc. This result is assisted by investigations of the FOA of the vapor and solid phase above real buried landmines in Cambodia and Bosnia-Herzegovina [9]. They did not find TNT in any case, but they found 2.4 DNT, 2.6-DNT and amino-DNT in Cambodia only. That means if dogs are able to detect mines and if no TNT could be found in the vapor phase above the mines what do they smell. Apparently they did not smell TNT. Today it is generally accepted that dogs do not smell the pure, single substance but an odor composition. It is accepted that dogs are successful with this strategy therefore it should be asked if that strategy would be more successful for sensor systems than to focus R&D-works on the pure explosives. In the case of TNT, RDX, HMX and others the vapor pressures of the pure explosive is mostly much smaller than the vapor pressures of byproducts or impurities. Therefore it might be more successful to identify these byproducts and to develop sensor systems which feature a very high sensitivity for these substances or odor bouquets composed of these substances.

WHAT ARE THE CONSEQUENCES FOR SENSOR DEVELOPMENT OR APPLICATION?

Today intensive work is conducted in research and development of sensors for the detection of explosives from the vapor phase. *Tab. 4* gives a rough overview about different methods and lower detection limits. All systems are under development and in each case it is propagated to enhance the sensitivity of the special system. But the question arises if it is necessary to enhance the sensitivity. The mentioned results about the vapor composition above mines show that in these cases no TNT was found. This gives a strong hint that the sensitivity of sensors in these cases will not influence the probability of the mine detection. A sensor which is highly selective for the detection of TNT will not be able to find mines under these circumstances. On the other hand dogs are able to find them. This observation gives two important inputs: Developer of sensors and dog handlers should come in closer contact to compare their special notes. And the second point might be that under the assumption that dogs are apparently able to detect mines, pyrotechnics etc. the detection strategy of dogs is more successful than the todays technical solution. That means the underlying problem should be reconsidered. Is it really the problem to detect explosives with very high purity or is it the problem that explosives have to be detected that are technical grade or home made? If the used explosives, independent of their chemical composition, are of technical grade they always contain byproducts or impurities. Mostly these substances have a higher vapor pressure compared to the pure explosive alone. It is very likely that dogs do not smell the pure explosives but some other substances of the odor bouquet of the explosive. Therefore it might be possible to combine different sensors which are highly sensitive for the different substances of the odor bouquet of the explosives, e.g. DNT or DNB in the case of TNT.

Tab.4: Lower detection limits for TNT of different systems (all systems under development)

fluorescence	1 pg/L	0,1 ppt
Antibodies	approx.: 1 pg after sampling	0,1 ppt
IMS	approx.: 50-100 pg/L	5-10 ppt
SAW	approx.: 100 pg/L	10 ppt
conducting polymers	200-400 pg/L	20-40 ppt
electrochemical	approx.: 300 pg/L	30 ppt
μ-electron capture detector	1 ng/L	100 ppt
airport sniffers	20 ng/L	2000 ppt

As a very early test experiments were conducted to discriminate pure TNT and two different TNT-containing explosives from the vapor phase. At first we used eight MOS-sensors for this discrimination and we expected that discrimination would not be possible if the TNT is the only substance which is detected. *Fig. 1* shows the result of these experiments. As can be seen only the explosive B could be slightly discriminated from the blind probe (synthetical air). In the next experiment we combined our electrochemical sensor, which shows actually a lower detection limit of 34 ppt for TNT, with the MOS sensors. *Fig. 2* shows the result of this experiment. Obviously the explosives show different patterns and could be clearly discriminated from the blind probe. Unexpectedly the probe B showed a slightly different pattern than pure TNT and probe B. In this case the electrochemical signal was analyzed at a fixed value so that we measured only different TNT concentrations. It was expected that if only a different TNT concentration leads to discrimination the pattern should be close together. This might hold for TNT and probe A but probe B showed a different pattern. Therefore it can be concluded from this very first experiments that, unattached by their sensitivity, the MOS deliver a different and additional aspect of explosive B.

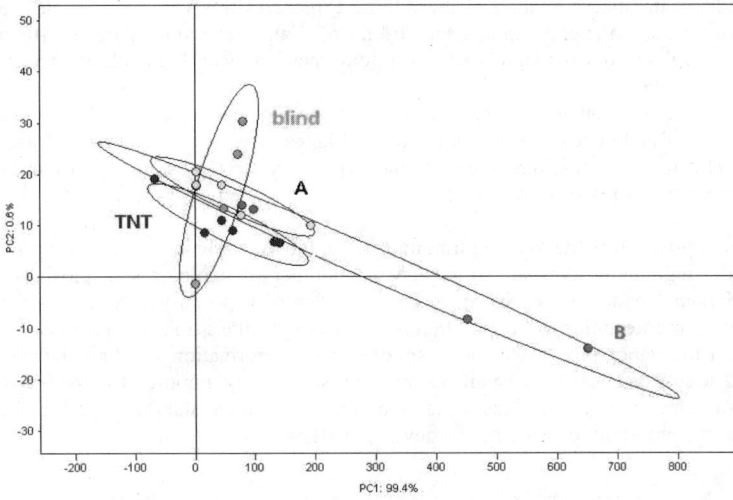

Fig. 1: Principal component analysis (PCA) on three vapor phases of different TNT containing explosive mixtures (eight metal oxide sensors)

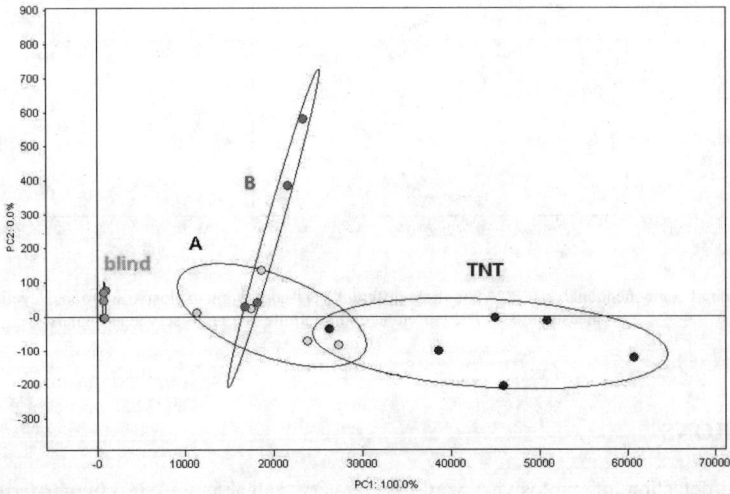

Fig. 2: Principal component analysis (PCA) on three vapor phases of different TNT containing explosive mixtures (eight metal oxide sensors combined with the signal of an electrochemical sensor)

It was unexpected that the discrimination of these three explosives was very easy by electrochemical methods in a liquid solution alone. As can be seen from *Fig. 3* all three probes show a different pattern if eight electrochemical sensors are simulated during one measurement. Presently we try to combine this method with the MOS.

Pattern recognition might be an interesting way for explosive detection, especially if the system is combined of sensors of different types which show the highest sensitivity for a special aspect of the explosives odor. This route would come closer to the dog's nose strategy as the use of single sensors which are highly sensitive for one substance only.

A more challenging problem is the very high number of explosives which could be used for terrorist attacks. Beside the high explosives an unmanageable number of explosives, explosives mixtures, pyrotechnics and home made explosives are to be considered. Especially mixtures like ANFO or inorganic explosives, pyrotechnics and in addition new explosives like TATP are a serious problem for the detection from the vapor phase. On the basis of present information our knowledge about the verifiability of these substances is very small. Therefore it seems to be important to focus R&D-works in these areas also. Especially in these cases the commutation between scientists, police and forensic scientists seems to be important to speed up the development of sensor systems.

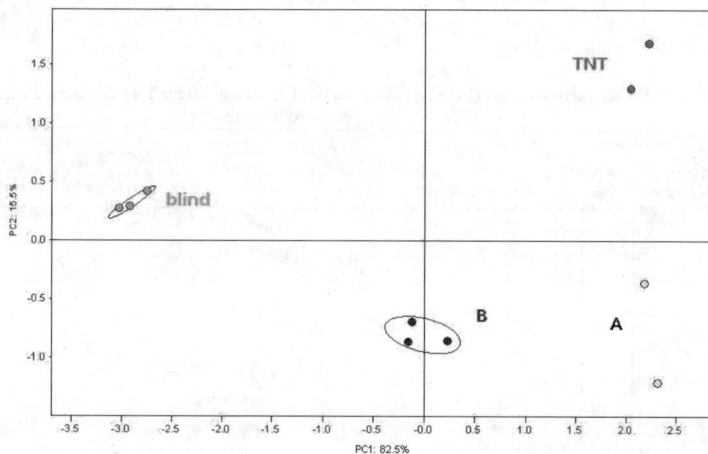

Fig. 3: Principal component analysis (PCA) on three different TNT containing explosive solutions, cyclic voltammetric measurement. The current at eight definite potentials served as signals.

CONCLUSIONS

Vapor phase detection of explosives would be a very valuable tool in counter-terrorism and in protection of society. Especially the search of large areas and the personal control would be applications which round off airport and station security and the supervision of public buildings and crowds of humans.

Several factors hamper the wide application of sensor systems for the vapor detection of explosives. One of the most serious problems concerning the explosive detection is the unmanageable number of possible chemicals and mixtures which could be used by terrorists. Beside high explosives numerous chemicals and mixtures could be used and in addition new explosives and detonators are developed.

At present the most serious substances seem to be TATP and HMTP, both are used by terrorists and the instruction for synthesis is available at the www. The problem of the detection of all different explosives is that the used chemicals differ in structure and therefore in chemical behavior, especially in the area of home made explosives. Therefore a specific explosive sensor which covers all possibilities seems to be implausible.

On the other hand trained dogs are today very successful in explosive detection. Up to now the knowledge about the odor bouquet which is identified by dogs as typical for explosives under realistic conditions is small. But this information would be a vital basis for the development of sensor systems which could be "trained" to smell completely different explosives under various environmental conditions. Maybe the strategy of chemical explosives detection should be reconsidered. The focus on the detection of the pure explosive might be changed in such a way that different chemical sensors are combined to detect the odor bouquet of explosives. It seems to be possible that such a system could be used for various applications and especially for various explosives, inorganic also, after a "training" period. Most important to reach this goal is that scientists of various disciplines and practitioners (police, military) and especially dog trainers and handlers come in close contact. Therefore NATO ARWs will be a very useful help to stimulate this discussion.

LITERATURE

[1] K. G. Furton, L. J. Myers, Talanta 54 (2001) 487
[2] R. L. Simmons, High-Impact Terrorism: Proceedings of a Russian-American Workshop, The National Academy of Sciences, 2002
     http://www.nap.edu/openbook/0309082706/html/171.html
[3] www.isis-online.org/publications/ terrorism/images.html
[4] http://www.totse.com/en/bad_ideas/ka_fucking_boom/
[5] http://www.licensed4fun.com/anarchist1.htm
[6] http://www.vetmed.auburn.edu/ibds/doglab.htm
[7] http://pubs.acs.org/hotartcl/cenear/970929/dog.html
[8] http://www.vetmed.auburn.edu/ibds/chemistry_laboratory/images/
     landmine_chromatogram.gif
[9] A.H. Kjellström, L.M. Sarholm, "Analysis of TNT and related compounds in vapor and solid phase in different types of soil",
     in Detection and Remediation Technologies for mines and minelike targets V, Proceedings of SPIE Vol. 4038 (2000)

METHODS OF DETECTION OF VAPORS AND TRACES OF EXPLOSIVES: A MODERN CONDITION AND PERSPECTIVES OF RESEARCHES AND DEVELOPMENT

A.A. REZNEV

SMC "Technological Center" MIET, 103498 Moscow, K-498, MIET, Russia

Modern methods for the detection and neutralization of explosives are important components of the complex actions for prevention of terrorist acts. Creation of effective efforts for detection of explosive devices now is one of the actual tasks of strengthening the activity of antiterrorist divisions. During development and creations of these means give the basic attention to detection of the various explosives making of a basis of any explosive device. Special attention to means and methods of explosive detection, especially to the vapors and traces, is given for the reason, that the last are unique direct attributes of the presence of explosives at suspicious objects.

Detection of the small amounts of explosives and their vapors in an atmosphere carried out with the help of an equipment realizing the following known analytical methods for detection of molecules of required substances in analyzed tests:

- Ion mobility spectroscopy;
- Gas chromatography;
- Mass-spectrometry.

Quality of the given analytical equipment is accepted for characterizing the value of sensitivity as which understand specific amount of the explosive contained in test found in the given signal / noise ratio in the analytical block. For the listed methods of the analysis sensitivity of the appropriate equipment makes about 10^{-13} g/cm^3 for ion mobility spectroscopy and achieves 10^{-14}-10^{-16} g/cm^3 for gas chromatography and mass-spectrometry.

M. Krausa and A. A. Reznev (eds.),
Vapour and Trace Detection of Explosives for Anti-Terrorism Purposes, 11-15.
© 2004 *Kluwer Academic Publishers. Printed in the Netherlands.*

The final parameter for the efficiency of the equipment is the detecting probability for the existence of an explosive. Despite of significant progress in development of analytical means for explosive detection and achievement of high values for the sensitivity, these means do not give the exhaustive decision for the problem of explosives detection. Moreover, the question of a quantitative estimation of their efficiency from this point of view remains insufficiently investigated. Considering a problem of improving of means for explosives detection, including their traces and vapor, it is necessary to bring attention to a number of basic questions which adequate decision is a basis at definition of strategy and ways to increase the efficiency of means for the detection of explosives. These questions consist of the following:

- in what manner the parameter of sensitivity of the equipment of vapors or traces explosive detection corresponds with the detection probability detection for an explosive at the given restrictions for the period of decision making and set of external factors and conditions;
- in what tactical advantages and disadvantages of methods for vapor and trace explosives detection consist in comparison with X-ray, nuclear and other methods;
- what place and role of hardware methods for vapor and trace explosives detection in the complex control of contents of objects with use of set of various means;

The applicability of systems for the detection of explosives vapors and traces will be in dispute as long as the sensitivity under practical conditions, as above mentioned, does not fit the requirements.
By results of long-term work of the state special services on prevention of acts of terrorism it is possible to allocate two types of tactical tasks:

1. Revealing explosive devices on a place of realization probable terrorist act before their operation.
2. Revealing subversive and terrorist means at attempts of their delivery to a place of realization of planned acts of terrorism.

The first type of actions provides the inspection of various objects and subjects in a place assumed terrorist act. Depending on situation of vehicles, elements of building designs, subjects of furniture and an interior, suspicious convolutions and packing can be subject to inspection.
The second type of actions is connected to the organization of the control on specially created points of examination. It can be examination of transport and cargoes on customs terminals, examination of people and their hand luggage at pass to various establishments and places mass a congestion of people or check of air passengers of their hand luggage and luggage, check of items of mail and cargoes etc.
Demands on the equipment for use in each of the considered types of actions will be specific. For example at points of examination basically stationary engines, which are fast, should be used because each object should be inspected during some seconds. The efficiency of this systems should be as high as possible and the false alarm rate should be low.
For the inspection of places assumed for terrorist acts basically portable or mobile engines could be used. Although the operation under these circumstances is also effected by the time like in the first case, the required productivity of these systems could be lower. But the systems should allow to carry out the inspection of objects in remote places and at difficult configurations.

Let's consider within the framework of the specified operational and tactical tasks a place and a role of means for detection of vapors and traces of explosives.

Analyzing the today situation of the Russian developments for the analytical equipment intended for the detection of explosive vapors and traces, it is possible to note, that the greatest distribution was received with gas chromatography equipment and ion mobility spectroscopy equipment. The most known samples of this engines are given in *Tab. 1*. Their basic technical parameters here are specified. From the table it is visible, that all devices have the high sensitivity close to TWT concentrations sated TNT vapors under normal conditions.

Tab. 1:

Type, model	Explosive	Sensitivity g/cm^3	Detection time (seconds)	Mass
Ion mobility spectrometer ("Shelf-DS")	Dynamite, TNT, EGDN, NG	10^{-13}	1-2	1,4 kg
Ion mobility spectrometer "M-02"	Dynamite, TNT, EGDN, NG	10^{-13}	1-2	1,3 kg plus power supplies
Chromatograph ("Echo-M")	Dynamite, TNT, EGDN, NG, RDX, PETN	10^{-14}	Collection 15-30 Analysis 30	

Considering productivity and efficiency of other types of the equipment for the decision of first of the mentioned tasks, it is necessary to recognize, that the gas-analytical equipment is the most suitable for realization of the express control of suspicious subjects in this task.

On the other hand, probability characteristics of explosive detection in real objects on the submitted means are absent. There are no data for the equipment of similar purpose of the western manufacturers.

At the same time, despite lacking such characteristics, from experience of practical work it is possible to formulate the basic tactical principles of application of means in a considered operational and tactical task. As it was already mentioned, from an available complex of means specified the gas-analytic equipment is the most productive. It is necessary to note, that productivity of inspection of object's with the help of this equipment is close to productivity of inspection with the help of manual metal detector. However, the information of the control by virtue of the features marked above is much higher. As time restrictions in the first task are less essential, than in a task of the control of a stream of objects, means of the analysis of vapors and traces of explosives can have rather high values of probabilities of the undetection and a false alarm. Thus if as a result of inspection it is made a decision on presence of vapors or traces of explosives, the given object is isolated and exposed to destruction. If it is made a decision on absence of explosives, inspection of object with the help of other, less productive means proceeds. Further the decision on presence or absence of an explosive is accepted on set of results of the analysis of various informative attributes of object. Now to state a quantitative estimation of acceptable values of probabilities of correct detection and false alarm it is represented inconvenient. In the long term these aspects of researches in a direction of improving the analytical equipment demand the greater attention and the appropriate support.

Let's come to the analysis of places and the challenge of the systems for explosives vapor and trace detection as above mentioned in the second task for the control of luggage of air passengers from the point of methodology view of complex inspection.

Now in the majority of the large airports the wide application was received with a method consecutive multistage the control of conveyor type. The principle of equipments combining consists that means settle down in a chain of the control in process of reduction of their productivity. Thus the decision on presence or absence of an explosive in controllable object's is accepted at each stage of the control, and the following stage carries out functions of additional check.

At such approach the responsibility for the accepted decision on absence of an explosive in unit of luggage grows in process of increase of a serial number of a stage of the control. At conveyor procedure of processing of a stream of luggage of means of detection of vapors and traces of explosives as against the first task appear less productive and, hence, settle down at the final stage of the hardware control. Therefore the probability of detection of an explosive on the basis of presence of it's vapors and traces in this case should be highest of all means used in a chain of the control.

In real conditions the concentration of vapors near to the object is some orders of magnitude lower than elasticity of the sated vapors for any type of explosive. Therefore to find out vapors of explosive in gas test it is possible only at its preliminary concentrating with the help of special devices or in the event that test is selected in immediate proximity from a surface on which there are traces of the explosive. Ways to increase the concentration of vapors, based on the use of closed volumes, result in an essential increase of inspection times, however, to present time of convincing proofs of their efficiency are not present.

The considered features of applications of means for detection of vapors and traces in practical conditions (situations) allow to formulate the basic directions and tasks for their improving.

In view of the reflected aspects of a modern condition of the equipment of the analysis of vapors and traces and its application in real conditions the basic directions can consist in the following:

- Increase of sensitivity of the analytical equipment;
- Perfection of ways and means for test collection and transportations of test the analytical block;
- Improvement of operational properties of the analytical equipment;
- Methodical maintenance of a problem of explosive detection with the help of analytical equipments;

New direction's for the essential increase of the sensitivity of systems for the detection of explosives is based on the development of small-sized stationary and mobile mass-spectrometry equipments. Sensitive mass-spectrometry equipments for the detection of vapors reach up to 10^{-16} g/cm^3 TNT that is much higher, than at any other known detector. Modern mass-spectrometry equipment is difficult in operation, demands for attendants of high qualification. These are principal causes which the wider hindered chemische applications of mass-spectrometry equipment for detection of vapors of explosive. At the same time, it is necessary to note, that the given direction now demands significant efforts in creation necessary scientific and technical bases for its realization.

More advanced are now Russian methods to increase the sensitivity by selective ionization of gas samples. Selective ionization is a perspective way to increase the sensitivity of both chromatography, and ion mobility spectroscopy equipments for detection of vapors.

Low efficiency of the analytical equipment at the decision of practical tasks on search of explosive promoted the further development under aegis of the United Nations and the acceptance of the International. Convention on marking plastic explosives.

Markers of plastic explosives show a vapor pressure which is by some orders of magnitude higher than the pure explosive. Because of their higher vapor pressure they create a higher vapor concentration near an object and there fore they can be found more easy by the detectors. Thus sensitivity of an analytical method as though raises.

However work on manufacturing of marking plastic explosives and development of analytical instruments which are sensitive for the tagging substances is only less intense.

As a possible way to increase the sample concentration, Russian experts developed the whirl wind vapor collecting system on the basis of a new sorption materials and elements.

One of the operational disadvantages of the modern equipment is that gas chromatography devices work only with super pure inert gases or nitrogen. It creates certain difficulties at the organization to operate the devices and leads to additional demands of material inputs.

Therefore one of the tasks which we put before ourselves, - to create the system working on atmospheric air without loss of sensitivity and with same weight and size parameters, as the equipment such as "Echo".

Considering, on the one hand, that fact, that unique direct attributes of presence at surveyed object of an explosive are traces or vapors and also taking into account the mentioned lacks of gas-analytical methods, it is necessary to recognize, that the task of detection of an explosive should be solved by complex methods with attraction of means of the analysis based on various informative attributes of an explosive.

In connection with the necessity of complex use of means we shall stop briefly on a question of optimum combining of means on an example of a task of the control of a stream of luggage and air passengers.

As against a principle of conjunction of the binary decisions, put in a basis of the accepted procedure of the control, on the given method at each stage of the control function of object's parameters is formed. Value of this function is compared to a threshold.

The physical sense of such method can be explained as follows. We shall consider a two-stage sequence of the control. In existing procedure of decision making if at the first stage the decision on presence of an explosive in luggage was accepted, the common decision will be positive if at the second stage the decision on presence of an explosive also was accepted, and negative if the decision on its absence was accepted. In optimum procedure decision on presence or absence of an explosive will be accepted depending on as far as is higher or below appropriate thresholds h1 and h2 there are values p1 and p2 target signals of the equipment of the first and second stages. Thus application of optimum procedure of decision making is most effective, when parameters of reliability and productivity of the equipment used at various stages, are close among themselves.

In this connection perspective way of increase of reliability of consecutive procedure of the control without decline of productivity now, in our opinion, is the use of QR-spectroscopy. The characteristics of the equipment of the given type achieved in Russia provide the same productivity of the control, as well as X-rays with two energies. Thus the reliability of the detection of plastic explosives, the most difficult for detecting by other kinds of the equipment, is highest of known means and methods.

The following characteristics for the detection such as RDX were achieved:

- Weight found out - 50 g;
- Probability of detection - more than 98 %;
- The probability of a false alarm - is not higher than 5 %;
- Volume of the analytical chamber - 200 dm^3;
- Time of the analysis - no more than 10 s;
- Realization of conveyor processing.

Comparative estimations of probability of detection of the explosive containing plastic EX's, show advantages of QR-spectroscopy in comparison with installation on thermal neutrons (TNA).

DETECTION OF EXPLOSIVES FOR TERRORIST-BOMBS AND LANDMINE CLEARANCE
DIFFERENT APPLICATIONS OF SIMILAR METHODS

HILTMAR SCHUBERT

Fraunhofer-Institut für Chemische Technologie (ICT), Joseph-von-Fraunhofer-Str. 7, 76327 Pfinztal, Germany; Tel.: 0049-721-4640 112; e-mail: hiltmar.Schubert@ict.fhg.de

INTRODUCTION

Since the development of modern analysis and detection methods supported by electronic means several different methods are available, which have been improved within the last decades with regard to precision, reliability, quickness and minimum test sample volume. These developments enable us to analyse substances very quickly - in some cases also on-line. The question is, which methods are suitable for application in the field under conditions of mine- and/or terrorist bomb detection. The importance of "Humanitarian Demining" in the last decades aiming at removing millions of landmines in the third world initiated research and development in the industrial countries. National and international programs were started and institutions were founded to solve one of the largest problems of our days. Non-government organisations, the "NGOs" have done an outstanding work in demining. But up to now, the deminers are working with relatively simple devices with very low frequencies and efficiency: Prodding, Dogs and Metal Detection.

The reason for this situation is very simple: all these efforts are financed by humanitarian programmes sponsored by international organisations or governments of industrial countries. This financial help is caused by bad conscience and responsibility felt for the third world and given without any future legal obligation. These unstable circumstances prevent the development of a free and open market, because nobody wants to guarantee the payback of investments producing an expensive demining artificial device.

There will be a quite another situation in the fight against terrorism. The danger addresses all of us, and the authorities are requested to protect people. One of the dangers are assaults of terrorists by the use of explosives. These circumstances produce a demand of detection devices, and, therefore, a market for industry will be formed. The consequence: Sophisticated devices for special applications to detect explosives will be on the market!

M. Krausa and A. A. Reznev (eds.),
Vapour and Trace Detection of Explosives for Anti-Terrorism Purposes, 17-21.
© 2004 Kluwer Academic Publishers. Printed in the Netherlands.

SENSOR TECHNOLOGIES

In *Fig. 1* a list of possible detection technologies is shown with a comment of maturity, cost and complexity. These comments may be changed, if we gain other understandings during the workshops of this year. May be, we will also add new technologies to this list.

It is not my intention to go into detail, because this is the theme of our workshop and two or three others this year.

I would like to concentrate my talk on the different conditions, behaviour, design, composition and properties of these explosive charges, because I have been director of a Fraunhofer Institute (ICT) for over 30 years, dealing with all kinds of energetic materials, and being an explosive materials expert in many working groups about demining.

CONDITIONS OF DETECTION

DETECTION OF LANDMINES

Landmines, usually produced in an explosive factory with professional knowledge, have a relatively small explosive charge of 25 – 250g of TNT. Sometimes also PETN, RDX or other explosives with high performance are used, but more for anti tank mines. The shape of the mines has mostly a rounded design, the charge is in a case made of plastic, steel or sometimes of wood and will be hidden in the ground 5 – 25cm deep. Mines which are connected by trip wires are fixed above the ground. The initiator reacts by pressure or by drawing the trip wire and consists of primary explosive in a metal tube.

The detection will usually be carried out in a rather simple manner by a time consuming prodding. Most of the demining companies and NGOs are using also trained dogs. Training a dog costs about 2000 Euro, and additional cost of 2000 Euro per year is necessary for livelihood, if a good performance is aspired. In recent years, researchers came to an understanding that the dogs do not smell TNT. It will be more a bouquet of odour of different items. Dinitrotoluol (DNT) is only one example.

If the mines contain metals, a metal detector can be very useful, improved devices are able to detect mines with minimum metal content. If other metal parts are in the ground, more or less false alarms will be the rule.

In a more homogeneous ground, "under ground radar" can be very helpful. The detection of landmines has to be done remotely controlled or with stand-off devices. Very often vegetation has to be removed before the detection can start. This must be done with a cutter also remotely controlled or under protection.

To my knowledge no sophisticated devices for mine detection are used in practice.

During the detection of landmines a distinct area is marked. Therefore only the deminer is acting in such an area. Therefore the danger for other people is limited.

DETECTION OF EXPLOSIVES USED BY TERRORISTS

The spectrum of terrorist charges regarding size, shape, confinement, composition and environment is extremely different to landmines.

SIZE AND SHAPE

The size of a charge may be between less than 1kg up to 1t and more. The different shape of the charge is dependent on the application. Usually the charge has an initiator cap, which is in some cases home made and therefore very dangerous to handle.
Explosive materials which are only transported to another place are more difficult to detect. Plastic explosives can be transported for instance in small quantities and can be transformed later on in a larger charge.
Explosive material may look like any subject you may imagine: Examples are tooth paste, textile rugs and clothing, flower pots, tablets, books, etc.

CONFINEMENT

Charges my be used with a soft or strong confinement consisting of metal, plastic, wood, cardboard or any other material in different thickness. The stronger the confinement used is, the more effective will be the detonation effect, also belonging to fragments.

COMPOSITION OF THE EXPLOSIVES

We must admit, that terrorists have sufficient knowledge about the behaviour of explosives, how to handle the material and how to prepare explosive charges with the different possibilities of composition. Beside literature there is also an access for everybody to the Internet, where you find informations how to prepare explosives and recipes for terrorist usage.("Terrorist Handbook", "Black Book", "Anarchist Handbook", Home-made Detonators", etc.).

There are different sources to get explosive materials:

1. Military Explosives
    Under normal circumstances it is relatively difficult to have access to explosives like the relative powerful TNT, RDX, HMX, Nitropenta, Semtex, etc.

2. Commercial Explosives
    These are explosives for road-building, for quarries and mining. Mainly Ammoniumnitrate based composition with fuel oil and/or with Nitrocompounds are used. For high performance also Straight Dynamites or Gelatine Dynamites based on Nitrocellulose and liquid organic nitrates with absorbents.
    All these commercialized explosives are accessable in most countries only by special licenses. Certainly, in some countries regulations are sometimes handled very careless – as the Oklahoma disaster has shown.
    In all Ammoniumnitrate based explosives a strong booster charge of high explosives for the initiation is necessary.
    Ammoniumnitrate is also used in large amounts as a component in fertilizers. But additional other components like Ammoniumsulfate cause the non-explosive behaviour of these mixtures.

3. Commercial Substances Suitable for Explosives
    These materials for explosives with relatively low performance are in most cases freely available. Examples are Black Powder, Smokeless Powder and many kinds of fire-works. Though these substances do not detonate, in most cases the effects in further distance are still remarkable.

4. Improvised Explosives
    All chemical compounds can be used as components for explosives if the oxygen content in the compound is more ore less high enough for a combustion without air.
    Mainly molecules with functional groups like:
    -   NO_2, - $NH-NO_2$, - $O-NO_2$ and NO_3+ (Nitro-, Nitramine-, Nitroxy-compounds and Nitrates as salts) and Peroxides are used.
    Also mixtures of salts like Nitrates, Chlorates and mainly Perchlorates with organic substances like plastic materials, plastizisers, organic liquids, solvents, etc. can be used.
    These combinations are very numerous and will reach quite more than 100 possibilities.
    Explosive materials can be prepared in liquid, plastic, slurry and solid state. Very easy to prepare are liquid systems with very high performance. These systems are mixtures of nitric acid, kerosin or nitrobenzene, etc. and were used in World War II from the allied air forces known as "House-Crackers".

5. Primary Explosives
    To detonate explosives a detonator primer is necessary. The primer is a capsule made from copper or aluminium with a small pressed charge of a primary explosive like lead azid. The charge will be detonated by an ignition device which react by a relative weak shock or by an electric impulse. Therefore terrorist try to get these initiators by an illegal way. Some terror assaults fail not having suitable detonators. In most cases only professionals prepare primary explosives. For the initiation of low energy explosive charges an additional booster with high energetic material like Nitropenta or RDX is necessary.

6. Powder Trains
    Experiences have shown that terrorists also use powder trains to ignite combustible materials in large volumes. In most cases pyrotechnic material is used for this purpose. The event of September 11th has shown, that catastrophic disasters can occur also without powder trains.

ENVIRONMENT

In opposition to landmines which are hidden in the ground in relatively lowly populated areas, for instance in open fields, roads, buildings, ditches or frontgardens, etc. explosive charges of terrorists can be found everywhere, also in highly populated surroundings even to cause an effect as large as possible.
The detonation can be fired by the cap directly, by remote control or with delay.
Because of the different surroundings detection methods may be different between landmines and terrorist bombs.
Spectroscopic means can be used for landmines only by stand off reflection while terrorist objects may be inspected also by transmission.
To prevent terrorist attacks, a main task will be to detect the explosive devices during the underground transport. This can be managed with objects ready to initiate, or the explosive is handled without an initiator only for transport.
Terrorist bombs are relatively easy to detect, because there is a case or box filled with a homogeneous material equiped with an initiating cap fitted with an electric time fuse. Sometimes the explosive charge is screened by a camouflage.
For transport reasons the explosive material may have any shape adapted to the environment. Therefore, such materials are only to be detected by their chemical composition. The relative high density of the material may be a useful indication.

EXPLOSIVES USED AS DISPERSER

To distribute nuclear, chemical or biological agents, explosives are used also. If granades or missiles are filled with these materials, an explosive is used with relatively low energy not to destroy the material to be distributed. Otherwise, the agents are distributed by pyrotechnical means.

CONCLUSION

The global political situation will cause an increase of terrorist actions in future. To protect people against this danger, an increase of efforts fighting against terrorism is urgent, and detection of explosives is one of the key problems of antiterrorism.

Because explosive materials can be transported in any shape and design, sensors are necessary which analyse the chemical composition of the material. Alternative measures are dogs or means which are called artificial noses.

We should keep in mind that we will never overcome terrorism, if we fight only against the symptoms. We should think about the reasons why terrorism is coming up.

The truth will be sometimes very inconvenient.

LITERATUR

1. S. Zeman et M. Hanus
    "Improvised Explosives and their Abuse for Bomb Attacks"
    NATO ARW Proceedings, Prague, 1997, "Explosives Detection and Decontamination of the Environment", published by University of Pardubice (Czech Republic), Explosive Holding, Prague

2. Carl E. Baum
    "Detection and Identification of Visually Obscured Targets"
    Braun-Brumfield, Ann Arbor, Mi 1998, ISBN 1-56032-533-X (case)

3. Hiltmar Schubert et Andrey Kuznetsov (Editors)
    "Detection of Explosive and Landmines"
    Proceedings, NATO ARW St. Petersburg,
    NATO Science Series II Vol. 66 Kluwe Academic Publishers 2001,
    ISBN 1-4020-0692-6 (HB) / ISBN 1-4020-0693-4 (PB)

4. NATO ARW "Advanced Research and Technologies for Detection and Destruction"
    Of Burried/Hidden, Anti Personal
    Landmines, Moscow, 1997 (not published)

5. International Symposium on Analysis and Detection of Explosives (ISADE)
    $1^{st} - 7^{th}$ (1983 – 2001) – every 3 years

VAPOUR AND TRACE DETECTION OF EXPLOSIVES

PETR MOSTAK

Research Institute of Industrial Chemistry, Explosia, 53217 Pardubice-Semtin
Czech Republic

ABSTRACT

Vapour detection and trace detection of explosives are the two basic methods of explosive detection. The boundary conditions in vapour detection and trace detection are discussed. The main advantage of vapour detection is a quick and operational easy search of objects, persons and buildings by hand-held detectors. The substantial problem is the temperature dependence of explosive vapour pressure. The limits of vapour detection at low temperature are demonstrated and discussed.

The further increase of sensitivity of detectors and improvement of preconcentrator capacity is required.

The trace or particle detection is not so quick and operational suitable but this method is independent on temperature conditions. Usually, one particle found and inserted in a detector inlet is sufficient for positive detection. The synergy of simultaneous use of vapour and trace detection can substantially increase the effectiveness of explosive detection.

The marking of plastic explosives for detection and improving of vapour detection is discussed.

INTRODUCTION

Vapour and trace detection belong to procedures of explosive detection frequently used in the search of presence of hidden explosive charges, contaminated objects and persons and in further cases connected with checking of luggage and persons and in the forensic work.

The tools for vapour and trace detection are two main systems, electronic detectors and special trained dogs. In both cases the final product entering the analytic part of system are explosive vapours, the vapours in trace detection are obtained by thermal desorbtion of collected particles of explosives or by desorbtion of explosive adsorbed from vapour on the surface of dust particles.

M. Krausa and A. A. Reznev (eds.),
Vapour and Trace Detection of Explosives for Anti-Terrorism Purposes, 23-30.
© 2004 *Kluwer Academic Publishers. Printed in the Netherlands.*

The detection work of dog is more complicated. In the process of sniffing the dog is apparently able to gain the vapours from the surface of particles of explosive or dust particles covered on their surface by explosives using the stream of warm and wet air emitted from the nose in quick short intervals and sucking in back the air enriched by evolved explosive vapours.

The feasibility of the vapour and trace detection by electronic detectors is dependent on many factors, the important factors and their consequences are discussed in this paper.

VAPOUR DETECTION

There are many factors important for successful detection of the vapour of an explosive. The most important are the vapour pressure of explosive and the sensitivity of detector (electronic detector or dog). The further factors are the diffusion and effusion of vapours from hidden explosive charge resulting in emission flow of explosive vapour. Important is also the adsorption of vapour on surfaces and dissipation of vapours into surrounding of the object in which is explosive charge hidden. This dissipation is dependent on the temperature and airflow near to the surface of the object.

Some factors are fixed and their value can be exactly estimated, other factors are very dependent on conditions present at the individual object. The very important are the barrier properties of materials, which are surrounding the explosive charge; it means properties of the content of luggage, parcel or box containing the explosive charge and also the barrier properties of packing.

Vapour pressure of explosive

The vapour pressure of explosive is the physical chemical property, which is specific for any chemical compound. The substantial difference between vapour pressure of individual compounds is typical also for explosives. Some examples are demonstrated in the *Tab.1*.

Tab.1: Vapour pressures of main explosives at 25°C

Type of explosive	Vapour pressure (ppb)
RDX	0.006
PETN	0.018
TNT	9.4
NG	580
AN	12

The useful information, which makes easier to assess the feasibility of detection using electronic detector is to express the vapour pressure in the mass of explosive contained in some volume of air. In the *Tab.2* the vapour pressure of some explosives is presented and also the vapour pressure of some further compounds, which are present in explosives as by-products or impurities in ng/l.

Tab.2: Vapour pressure of some explosives and by-products at 25°C

Compound	Vapour pressure (ng/l)
RDX	0.04
PETN	0.09
HMX	0.38
TNT	70
NG	4000
2,4 DNT	1440
1,3 DNB	8140

Taking into account the sensitivity of up-to-date electronic detectors, which is in the range of 0.02 – 0.2ng, we can consider the feasibility of detection of some explosives as rather difficult. The low vapour pressure of RDX, PETN and HMX is the reason why the detection of these explosives is not easy and usually needs a preconcentration step which can be achieved by adsorption capacity of the membrane or sieve in the front part of detector. The separate preconcentration units having higher efficiency are also used.

The much higher vapour pressure of 2,4DNT and 1,3DNT in comparison with TNT is the reason, why TNT containing mines are detected easier by vapours of these by-products than by TNT vapours [1] [2].
In detection of mines in which TNT is used, it seems, that the dog is able to identify the integral "smell" of the mine rather than the vapour of TNT or further compound.

Influence of temperature on vapour pressure of explosives

The vapour pressure of chemical compounds is dependent on the temperature by logarithmic equation: $\log P = A - B \cdot T$
These relations experimentally estimated for main explosives are presented in the *Tab.3* [3].
The graphic demonstration of this dependence at TNT is shown in the *Fig.1*.

Tab.3: Dependence of vapour pressure of some explosives on temperature

Compound	Equation
RDX	$\log P \text{ (ppt)} = -6473/T(K) + 22.50$
PETN	$\log P \text{ (ppt)} = -7243/T(K) + 25.56$
TNT	$\log P \text{ (ppb)} = -5481/T(K) + 19.37$
NG	$\log P \text{ (ppb)} = -4602/T(K) + 18.21$
AN	$\log P \text{ (ppb)} = -3541/T(K) + 12.97$

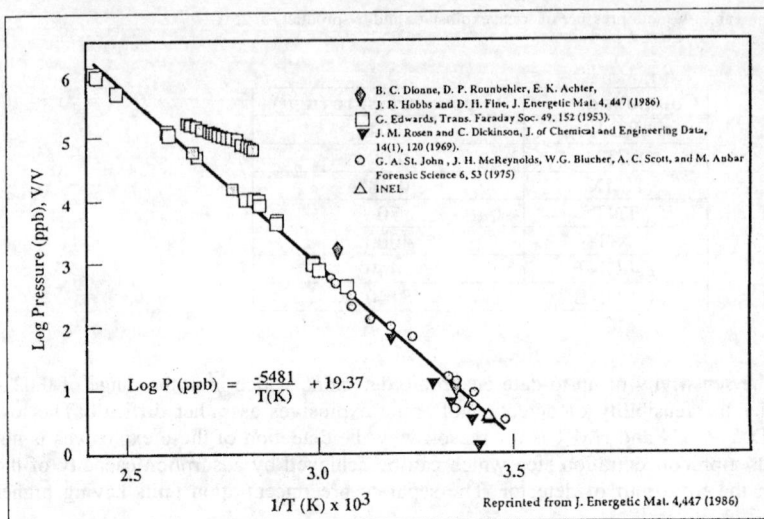

Log P (ppb) = $\dfrac{-5481}{T(K)}$ + 19.37

Reprinted from J. Energetic Mat. 4,447 (1986)

Fig.1: Dependence of the vapour pressure of TNT on temperature

The dramatic changes of the vapour pressure of some energetic materials in the temperature range 0 – 30°C can be seen in the *Tab.4.*

The vapour pressure changes expressed in multiples of the vapour pressure at temperature of the lower level of the temperature range, is presented in the *Tab.5.*

It can be seen that the dependence of vapour pressure on temperature is most profound at RDX at PETN, it means at explosives having the low vapour pressure. This fact brings some significant limit in detection feasibility of these explosives at low temperatures. The problem can be solved by effective preconcentration of explosive vapours before analytical sequence or by increasing the sensitivity of detectors.

Tab.4: Vapour pressure in the temperature range 0 – 30 °C

Vapour pressure	0°C	10°C	20°C	25°C	30°C
RDX (ppt)	0.06	0.43	2.57	6.03	14.45
PETN (ppt)	0.11	0.93	6.92	18.20	45.71
TNT (ppb)	0.19	1.00	4.57	9.55	19.05
NG (ppb)	22.39	89.12	316.2	588.8	1047.1
AN (ppb)	1.00	2.88	7.35	12.30	19.05

Tab.5 : Vapour pressure changes in multiples

Compound	0 – 20°C	10 – 20°C
RDX	41	6
PETN	63	7.5
TNT	24	4.5
NG	14	3.5
AN	8	2.6

The results presented above confirm, that the influence of temperature on detection is substantial, this temperature dependence is strongest at RDX and PETN. The decreased detection ability of vapour detection at lower temperatures is caused not only by low vapour pressure but also by further factors, which are influenced by temperature (diffusion of explosive vapours through barrier materials, emission flow).

TRACE DETECTION

The trace detection usually means the detection of contamination of surfaces by explosive particles, which are dispersed and fixed on various objects by manipulation or contact with explosives. Traces of explosives have the form of fingerprints made by hands contaminated at the preparation of explosive charge, also the individual explosive particles can be dispersed in the area of production or manipulation with explosives. Air movement transports such particles which contaminate surfaces in rooms, various objects and also clothes of persons .
The detectable contamination of person is caused sometimes by NG and PETN from drugs for cardiac in which are these compounds substantial components. Particles of explosives can be also formed by condensation of explosive vapours at temperature changes.
In some conditions, the fine dust particles can adsorb explosive vapours on its large surface. Such explosive traces can be in the process of detection evolved by the sampling used by electronic detector and also by dog.
The secondary trace contamination is often taking place when persons, luggage or other objects come in contact with contaminated surfaces.
The substantial difference between vapour detection and trace detection consists in the fact, that detection of explosive vapours is usually the sign of the presence of a hidden explosive charge. On the other hand, the positive detection of explosive traces in many cases means only some incidental contact with explosive or contaminated surfaces, even only contact with heart medicine.

Many publications have been dedicated to quantification of fingerprints on various surfaces [4, 5, 6]. An interesting conclusion was achieved, that especially plastic explosives are producing massive fingerprints on most surfaces. It was found, that also the manipulation and forming of plastic explosives is connected with an specific physical process which causes a substantial contamination by explosive particles in vicinity of such activity .
The detection of traces or particles by electronic detectors is relatively easy. The reason of this is the fact, that the amount of explosive in any trace is usually sufficient for comfortable detection.
One fingerprint of plastic explosive contains 500–3000ng of PETN or RDX. Considering the sensitivity of up-to-date detectors, which is in the range of 20–200pg, it is clear, that when we are able to gain even only 5% of the explosive contained in the fingerprint, we have much more explosive than we need for detection. Using standard collection methods (swabs or vacuum sampler), we can collect much more than 5% of the explosive present in the fingerprint.

Looking for the particles of explosives dispersed by an air movement, it can be estimated, that the collection of 1 particle is sufficient for the successful detection.

One 30 micron particle, which represents an average size of explosive particles, has the mass of approx. 40ng and this quantity is 100 times higher than the sensitivity of the electronic detectors.

The high effectiveness of the particle detection has been achieved in the walk-through portal detection. This system enable to detect the presence of explosives on the surface of person by washing the person by the air stream and collecting and analysing obtained particles and vapours. It was proved, that it is possible to detect the trace of the plastic explosive C-4 having the mass 100-500ng fixed on the surface of the person clothing [7].

SIMULTANEOUS VAPOUR AND PARTICLES DETECTION

The detectors, which are able to detect both explosive vapours and particles, can be more effective in the detection of hidden improvised explosive devices. In some cases we have the good chance to detect the explosive vapours emanated from hidden explosive charge, but we are not able to find the traces on the checked surfaces. In other situations we are able to find the traces but not the vapours.

Therefore, the simultaneous vapour and particles detection gives us the synergetic effect, which increases the probability of positive detection.

We are to take into account also the fact that each of the 2 methods of detection gives different information. Detecting traces of explosive (fingerprints, particles), we obtain the information, that the surfaces of checked objects or persons are contaminated by some unspecified contact with explosive or contaminated hands or by manipulation with explosive in vicinity of these objects. Thus, the detection of explosive traces is not the prove or strong suspicion that explosive charge is present, the detection can be signalled already in presence of nanogram mass of explosive. Such mass of explosive evolves not enough explosive vapours for vapour detection.

On the other hand when the explosive vapours are detected, we can deduce that the bigger mass of explosive is present and the probability, that the source of vapours is some explosive charge is relatively high.

It means, that for the good evaluation of detection results, it is important to know if traces, vapours or both were detected.

DETECTION OF MARKED PLASTIC EXPLOSIVES

Plastic explosives contain as the main explosive component RDX, PETN or mixture of these explosives. The vapour pressure of PETN and RDX is low and therefore, the detection of vapours is rather complicated. The detection of plastic explosives by detection of traces makes no problem and is easier than detection of traces of some other explosives.

Taking into consideration the cases, in which plastic explosives were used in bombing of civil planes, the UNO was decided to establish the marking of plastic explosives for detection, with the aim to increase significantly the feasibility of the effective vapour detection of plastic explosives used as explosive charge in bombing devices.

The relevant international convention was prepared by ICAO and the Convention on the Marking of Plastic Explosives for the Purpose of Detection was agreed during the International Conference in Montreal on March 1,1991.

The detection agents and their minimum concentration in the finished plastic explosives at the time of manufacture were specified in the Technical Annex to the Convention [8]. During the years in which some experience with marking for detection was gained, some changes in the Technical Annex has been agreed and the actual specification of detection agents are shown in the *Tab.6.*

Tab.6: Detection agents and minimum concentrations

Name of detection agent	Molecular formula	Minimum concentration
Ethyleneglycoldinitrate (EGDN)	$C_2 H_4(NO_3)_2$	0.2 % by mass
2,3-Dimethyl-2,3-dinitrobutane (DMNB)	$C_6 H_{12} NO_2)_2$	0.1 % by mass
para-Mononitrotoluene (p-MNT)	$C_7 H_7 NO_2$	0.5 % by mass

In the phase of proposal is the raising of minimum concentration of DMNB to 1%. This measure should ensure the presence of sufficient DMNB concentration during the whole shelf-life of plastic explosive. In last years some producers increased voluntarily the DMNB content to 1% to achieve this goal. The vapour pressures of some explosives and marking agents are demonstrated in the Tab.7.

Tab.7: Vapour pressures of explosives and marking agents at 25°C

Compound	Vapour pressure (ng/ml)
o-MNT	860
EGDN	320
p-MNT	170
DMNB	12
NG	4
TNT	0.07
RDX	0.04×10^{-3}
PETN	0.09×10^{-3}

The main effect of marking agent on detection is the high vapour pressure of marking agent, which enables to gain the sufficient mass of vapours by electronic detector for positive detection.
Considering the fact, that the vapour pressure of DMNB over marked plastic explosive is approx. 50% of the vapour pressure of pure DMNB, we can see that vapour pressure of DMNB is higher by 5 orders in comparison with RDX and TNT. It means, that the feasibility of the detection of marked plastic explosives is on the level of NG containing explosives at which the feasibility of detection by NG vapours is high.
In the Czech Republic all plastic explosives produced are marked for detection from 1991. There is also the intention to mark the old stocks of plastic explosives.

CONCLUSIONS

The improvement of the effectiveness of vapour and trace detection achieved in last years is high, but not sufficient in all situations.
The low detection effectiveness of vapour detection at decreased temperatures should be solved by higher sensitivity of detectors and by more effective preconcentrators, having the preconcentration factor 1000 at minimum.
The synergy of the compact system combining both particles and vapour detection can be substantial in some cases.
The marking of plastic explosives for detection is an effective way how to increase the feasibility of vapour detection of these explosives
The most effective detection system should combine the vapour and particles detection with bulk detection.

REFERENCES

[1] M.Fisher, Detection of Trace Concentrations of Vapor Phase Nitroaromatic Explosives by Fluorescence Quenching of Novel, Polymer Materials, Proceedings of the 7th International Symposium on the Analysis and Detection of Explosives, June 25-28, 2001 Edinburgh, Scotland

[2] M.Krausa, H Massong, P.Rabenecker, H.Ziegler, Chemical Methods for the Detection of Mines and Explosives, Proceedings of the NATO ARW on Detection of Explosives and Landmines, September 9-14 2001 Saint Petersburg, Russia

[3] B.C. Dionne, D.P.Rounbehler, E.K.Achter, J.R.Hobbs, D.H.Fine, Vapour Pressure of Explosives, Journal of Energetic Materials, 4, 447-472 (1986)

[4] P.Neudorfl, M.A.Mc Cooeye, L.Elias, Testing Protocol for Surface Testing Detectors, Proceedings of the 4th International Symposium on Analysis and Detection of Explosives, September 7-10, 1992 Jerusalem, Israel

[5] L.Elias, Swab Sampling of PETN and RDX Deposits on Various Surfaces, 4th Workshop on the detection of explosives, Ascot, Berkshire, UK

[6] P.Mostak, M.Stancl, Detection of Semtex Plastic Explosives, Proceedings of the NATO ARW on Detection of Explosives and Landmines, September 9-14 2001 Saint Petersburg, Russia

[7] F.Kuja, et al, A Walk-Through Portal for Trace Detection of Explosive Particles and Vapours, Proceedings of the 7th International Symposium on the Analysis and Detection of Explosives, June 25-28, 2001 Edinburgh, Scotland

[8] Convention on the Marking of Plastic Explosives for the Purpose of Detection, ICAO Montreal 1991

ANALYSIS OF EXPLOSIVE VAPOUR EMISSION TO GUIDE THE DEVELOPMENT OF VAPOUR DETECTORS

S.R. DIXON, D.M. GROVES*, P.A. CARTWRIGHT, S.N. CAIRNS, M.D. BROOKES AND S. NICKLIN
* Author for correspondence. Tel 01959 892448, fax: 01959 892506
email DMGroves@dstl.gov.uk

Detection Department, Defence Science and Technology Laboratory,
Dstl Fort Halstead, Sevenoaks, Kent. TN14 7BP

ABSTRACT

Canine olfaction is an established technique for the detection of explosives. Development of sensitive vapour detection instrumentation is currently underway worldwide to provide a similar capability. It would greatly aid this development if the mechanism by which dogs achieve explosives detection were more fully understood.

The specific aims of this programme are to determine whether canines are successful in detecting explosives because:

♦ the high sensitivity of the biological system enables dogs to specifically recognise a scent picture containing vapour from low volatility explosives, or
♦ whether dogs alarm on an incomplete scent picture consisting of a mix of the more volatile vapours from characteristic materials present in plastic explosive.

To achieve these aims, a set of specially manufactured RDX based plastic explosive samples have been prepared, where each of the samples lacks one component. These samples were then analysed to establish the components present and their concentrations. Once complete, canines initially trained to detect samples of the whole explosive were exposed in double blind trials to vapours from each of the depleted samples and their responses recorded. Correlation of the canine and instrumental results allows deduction of the relative importance of the different ingredients used by the dog for detection. The first phase of this programme: the analysis of vapours of an RDX-based plastic explosive and its constituents, and initial canine training has been reported previously [1]. This paper describes the completion of the work and in particular the results of the canine analysis and correlation of the headspace analysis and canine detection rates. Conclusions are drawn and the implication for the design of future explosive detectors discussed. For the convenience of the reader, part 1 is reproduced again here in full.

31

M. Krausa and A. A. Reznev (eds.),
Vapour and Trace Detection of Explosives for Anti-Terrorism Purposes, 31-42.
© 2004 Kluwer Academic Publishers. Printed in the Netherlands.

INTRODUCTION

Canine olfaction has demonstrated its utility as an effective tool for the detection of explosives and other contraband. A reliable method of remote air sampling followed by interrogation of the sample by olfaction has also been developed and used with success. These techniques demonstrate the important role that sensitive vapour detection currently plays in explosive detection. Canine olfaction does, however, have some drawbacks. Dogs can suffer periods of poor health and lapses in concentration. Both of the above affect their ability to respond to target vapours and both are difficult to monitor.

Development of sensitive vapour detection instrumentation is currently underway worldwide [2], which aims to provide a similar capability to that provided by canine olfaction, but with the following inherent advantages:

♦ detection combined with identification of a range of target materials

♦ quantification

♦ ability to verify equipment performance.

Vapour detection of plastic explosives by instrumentation can potentially be achieved through one of two approaches:

♦ extremely sensitive specific instrumentation capable of detecting vapours from the explosive components themselves

♦ less sensitive broad range instrumentation capable of detecting the associated non-explosive volatile components.

The first approach represents the ideal. However detection of plastic explosives by the pure explosive component is extremely difficult due to their very low vapour pressures. This will require extremely sensitive equipment, capable of detection of explosive vapours at many orders of magnitude below their saturated vapour pressures.

The second approach requires less sensitive equipment, capable of detecting a wide range of organic hydrocarbons. Such equipment will require parallel detection and pattern recognition techniques to provide a real time capability.

It would greatly aid the decision process regarding which strategy to take if the mechanism by which canines achieve explosive detection were fully understood.

RESEARCH OBJECTIVES

The overall objective of the research is to determine the mechanisms of canine olfaction and to utilise this to underpin the development of future vapour detection equipment. The specific question the research aims to address is:

♦ whether canine olfaction is successful because the high sensitivity of the biological system enables dogs to recognise and alarm on vapours from the actual explosives themselves, which in the case of RDX and PETN will be present at extremely low levels;

or,

♦ whether dogs alarm on a scent picture consisting of a mix of the more volatile vapours from the characteristic non-explosive ingredients present in plastic explosives.

STRATEGY

To achieve these aims outlined above, a set of specially manufactured samples of an RDX based plastic explosive have been prepared, where each of samples lacks one or more components. These samples were then analysed to establish the components present in the headspace and their relative concentrations. Special detection canines initially trained to detect samples of the whole explosive were exposed in double blind trials to vapours from each of the depleted samples and their responses recorded. Correlation of the canine and instrumental results allows deduction of the relative importance of the different components used by the dog for detection.

EXPERIMENTAL

Manufacture of samples

Samples of Rowanex 4100 were prepared in collaboration with BAE Systems. Two hemispherical Harvey Pan mixers, each of capacity 2.5L, were used in manufacture of all samples. Both units were identical, except in design of the rotating blades. The first used an anchor blade and the second a butterfly blade. Each pan was maintained at the designated temperature of 50°C to 100°C by heated oil circulating within a cavity in the pan wall. Each machine was housed within an individual bay contained within an explosive processing building. Ingredients were added manually at various stages of the manufacturing process, when the mixer was stationary. For reasons of safety, mixing of all compositions containing explosives was performed remotely from outside the mixing bays, under guidance of CCTV cameras situated inside the bay.

Contamination control

To ensure success of this work, it was vital that there was no cross contamination between components of the various mixtures, in particular the transfer of explosives to non-explosive containing mixtures. Precautions were taken to safeguard against contamination by both vapour and particulate transfer mechanisms at all stages of the process including manufacture, handling or storage of the samples.

Contamination control during manufacture of samples

Both the insides and outsides of both pan mixers used were scrupulously cleansed by the following procedures, applied in sequence:

♦ Scrubbing using solutions of detergent.

♦ Rinsing with clean water.

♦ Purging by filling up the pan mixer with fresh binder followed by stirring and decanting to waste.

Confirmation of the effectiveness of the cleaning process in removing explosive contamination was obtained by dry swabbing the interior of both pan mixer bowls and blades using Sharkskin filters. These filters were subsequently analysed using a calibrated explosives particle detector (Itemizer, GE Ion Track), capable of detection at trace levels. The cleaning process typically required many time-consuming repetitions of the above sequence until three successive analyses revealed both pan mixers were completely free of any detectable explosive contamination.

The design of the manufacturing sequence was such that preparation of all non-explosives mixtures was completed, before preparation of mixtures containing explosives commenced.

Contamination control during storage

Each mixture was manufactured in sufficient quantities to enable preparation of ten samples. The samples were decanted from the pan mixer into glass reagent bottles. Individual sample packaging was completed by sealing the bottle-cap join with a layer of Parafilm and double wrapping the bottle using two polythene bags, each bag heat sealed shut.

The ten samples of each mixture, which formed a complete set, were packaged into a cardboard box (30cm x 17cm x 13 cm), the lid and edges of which were sealed using tape. A further barrier to both vapour and particulate contamination was obtained by placing this box within a large polythene bag. Finally each batch of ten samples was placed in a fibreboard box along with similar sets of samples. Batches of non-explosive samples were packaged together and kept separate from explosive containing batches.

Sample handling

When removal of non-explosive samples from storage was required, the entire box was moved to a trace laboratory (explosives free) whereupon only the requisite number of bottles were removed for analysis or use in canine olfaction trials. The packages were then resealed and the box returned to the store. Samples were removed from the double wrapped glass bottles in the laboratory prior to dispensing into ceramic boats or conical flasks for headspace measurements.

All batches of explosive samples were necessarily stored in an explosives magazine. When samples were required, the entire box was removed to a processing area where small samples were removed from selected bottles and transferred to new vials for laboratory analysis. The sample bottles were sealed and repackaged for return to the magazine.

Sample analysis

The primary analytical technique used was headspace GC-MS analysis, performed by the complementary techniques of thermal desorption and solid phase micro-extraction. Thermal desorption GC-MS, whilst having the advantage of high sensitivity conferred by the preconcentration effects inherent in the sample collection process, is limited in that only relatively volatile compounds will pass through current TD units and artefacts may be generated by the adsorbent materials. Thus headspace measurements were also conducted using solid phase micro-extraction GC-MS (SPME-GC-MS) whereby the fibres used to collect vapours are inserted directly into the liquid injection port thereby overcoming limitations of thermal desorption equipment. Also any artefacts produced might be expected to be different to those associated with adsorbent materials.

Headspace analyses were supplemented wherever possible by impurity analysis of the bulk material by liquid injection into the GC-MS. In this technique, pure material was dissolved in solvent to produce a concentrated solution then injected into the liquid injection port of the instrument. Although the amount of pure component introduced greatly exceeded column capacity, any trace impurities present should be within column capacity.

Instrumentation

Analysis of all materials was performed using an Agilent GC-MS system, consisting of gas chromatograph model 6890 fitted with electronic pressure control (EPC) coupled to a 5973N mass selective detector, operated with electron impact (EI), and positive ion chemical ionisation (PICI). A 60m column (0.25mm with 0.25µm ZB5 stationary phase) was used in conjunction with a low temperature ramp rate to enable resolution and accurate integration of most components derived from the binder materials within 75 minutes. Analysis of the volatile impurities associated with the anti-oxidant was also possible under the same conditions with retention times between 39mins to 55mins. The pure material could also be chromatographed, but appeared during subsequent analytical runs. All headspace studies and impurity analyses utilised this column except for RDX, which decomposed irrespective of mass injected or temperature program employed. Thus a shorter column (15m x 0.25mm with 0.25µm ZB5 stationary phase) was used for this material using positive ion chemical ionisation where necessary to increase intensity of molecular ions.

Thermal desorption was performed using an automated thermal desorber (Perkin-Elmer ATD400) with the transfer line coupled directly to the analytical GC column by a press fit connector. All instrument parameters aside from thermal desorption parameters were computer controlled via HP Chemstation software. The MS was tuned in electron impact mode using perfluro-tributyl-amine to produce a high mass peak (m/z 502) with an intensity 10% of base peak ion (BPI). Tuning parameters remained unchanged throughout the analyses except for the electron multiplier voltage. This was progressively increased to maintain a constant response to a benzoquinone standard (10ng) spiked on to a blank tube positioned at the start of each queue of tubes analysed.

Selected thermal desorption studies were repeated using positive ion chemical ionisation, using methane as reagent gas, to enhance any high mass information obtained in fragmentation patterns resulting from many of the materials derived from the binder components.

Headspace sampling method

A dynamic headspace technique was employed, whereby samples were placed into clean ceramic boats (85mm long x 12mm wide x 8mm, deep) which were placed into a home built vapour generation oven. Sufficient material was weighed into the boat so that sample came up to the lip of the vessel to provide a relatively constant surface area to the purging gas.

For each raw ingredient and mixture, three samples were taken. For raw ingredients, three samples were taken from the same batch. For mixtures, samples from the 1st, 5th and 10th bottles were removed. Each sample was introduced in the oven, and after an equilibration period of two hours, the vapours were collected on Tenax packed tubes for 20mins, collecting 2000ml of vapour.

After removal of the sample, the oven was left for 1 hour to purge the sample chamber of any remaining volatiles. Confirmation of the effectiveness of purging was obtained by collecting the emissions for subsequent analysis before a fresh sample was introduced into the oven.

Impurity analysis

Solvent blanks were prepared by decanting a small portion of neat solvent into a sample vial (acetone HPLC grade). Concentrated solutions of solid compounds were prepared by dissolving pure material (1g) in this solvent (10ml) to give a highly concentrated solution (10.0% w/v) equivalent to 100,000 ng/μl). Liquid samples were analysed neat. Analysis of the solvent blank was performed and providing it was pure, analysis of the concentrated solution followed. All impurity analyses were performed by a liquid auto-injector with repeated washing of the syringe after injection of a concentrated solution. A splitless injection technique was employed and whilst concentration of the pure material introduced on column greatly exceeded column capacity, any impurities were only present at trace levels and produce symmetrical gaussian shaped peaks.

Canine olfaction

Preparation of training samples

Samples for canine training were prepared by depositing small samples of bulk materials onto filter papers. Fresh samples were provided each week to eliminate potential contamination problems and drying out effects. All materials used, with the exception of explosives, were stored before use in a dedicated clean laboratory.

Positive samples were prepared by depositing a small quantity of the full explosive composition (50-70mg) on to a filter paper (Whatman No. 41) which was then folded over and inserted into a Universal tube, (polycarbonate body, polypropylene cap). Each tube was then inserted into an aluminised anti-static bag and closed by heat sealing.

Preparation of blank samples was performed entirely in the clean laboratory and was identical in every respect to the explosive samples, except that empty filter papers were substituted in place of explosive loaded filters. Each filter was placed into an individual universal tube then heat-sealed into an anti-static bag for transportation.

Background odours or interferences were introduced into the training regime to ensure the dogs were detecting the target odour as opposed to merely detecting differences between positive samples and blanks. Thus a range of interferences including the animal repellent Renardine, coffee, and musks were chosen which might present a graded challenge to the dog. All samples with the exception of Renardine were presented as positive explosive samples i.e. 50-70mg of material smeared on a filter paper. A small quantity of Renardine (0.1ml) was applied by pipette to the filter paper.

Canine training

Canine olfaction was conducted using two passive detection dogs with no previous experience or exposure to energetic materials. Training was carried out by the reinforcement method. Samples were placed onto stands, set out in a row approximately 0.6m apart. A typical training run involved placement of four blank/interference samples and one positive sample, the full explosive mixture. A successful canine detection was followed by a reward. The sequence was repeated a number of times, and at random intervals, blank runs containing no positive samples were introduced. Typical training sessions lasted from one to two hours per day and continued for a total of eight weeks.

RESULTS

Results of the vapour analysis

The results of the vapour analysis conducted are summarised below.

The vast majority of components detected within the vapour fingerprint of all mixtures, both explosive and non-explosive, were derived from the binder. A mass chromatogram of analysis of the pure material is shown below in *Fig.1*.

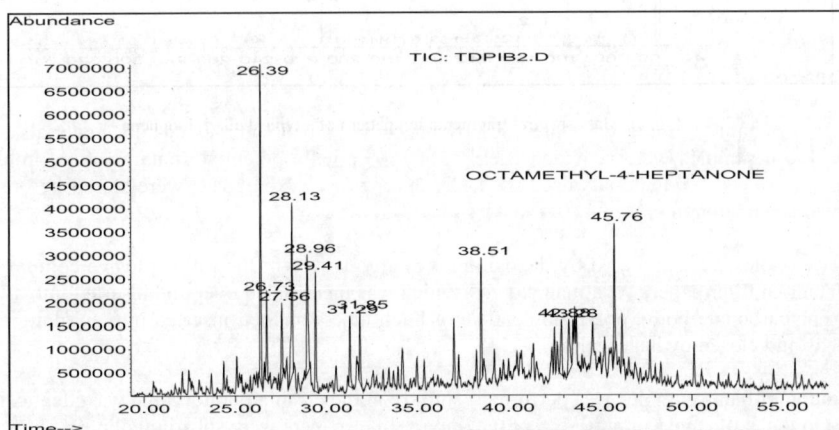

Fig. 1: Mass chromatogram obtained from analysis of binder

The majority of vapours emitted by this material were believed to be highly branched alkenes and alkanes, none of which produced detectable molecular ions in EI and PICI detection modes. A particularly intense peak (retention time 45.76mins), observed within the vapour fingerprint of all binder mixtures, was a compound which was consistently identified by library searching as octamethyl-4-heptanone. Whilst the library searched fit value for this compound was relatively low (Q=40 out of 100) this class of compound has been identified as an oxidation product in the extrusion Process of low density polymers at elevated temperatures. The fragmentation pattern observed was consistent with this identity. Hence it was believed the assignment was most likely correct. These compounds were believed to be significant contributors to odours arising from heated plastics [3].

Consistently detected within all mixtures, both explosive and non-explosive, were a homologous series of eight alkyl-2-thiophenes originating from the binder. They possessed a distinctive fragmentation pattern comprising two prominent peaks, the intense base peak ion due to the aromatic thiophene moiety and a relatively intense molecular ion. The mass spectral fragmentation pattern of a typical alkyl thiophene is shown below in *Fig. 2*.

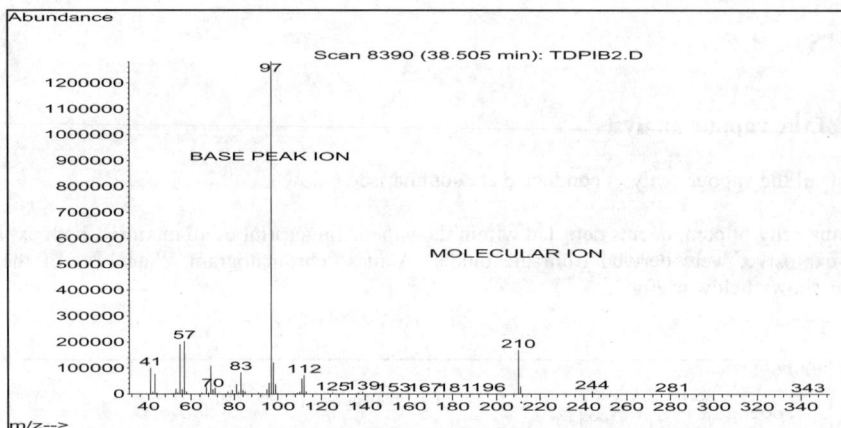

Fig. 2: Mass spectral fragmentation pattern of a typical alkyl thiophene

The vapour fingerprint of all explosive mixtures contained a prominent peak due to cyclohexanone which was used to recrystallise the RDX. Also identified as a smaller peak within some (but not all) of the explosive containing samples, was cyclohexanone-2-cyclohexylidene. TIC GC chromatograms obtained from the analyses of RDX are shown below in *Fig. 3*.

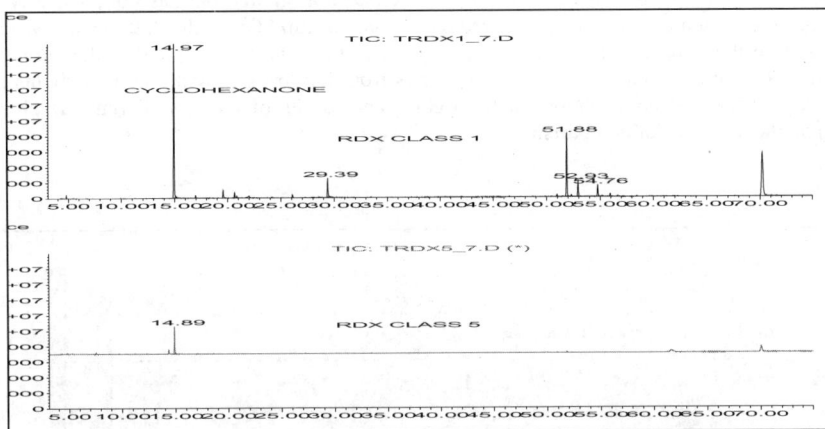

Fig. 3: GC chromatogram obtained from analyses of RDX

The vapour fingerprint of the anti-oxidant in the explosive as determined by TD-GC-MS comprised seven volatile impurities, which appeared towards the end of the analytical run. Four of these materials were identified with a high degree of certainty as alkyl substituted phenols. It was believed another two impurities were also phenolic in nature. Identity of the seventh remained uncertain. The pure anti-oxidant itself was not sufficiently volatile to be detected by the analytical technique employed. A chromatogram obtained from headspace analysis of anti-oxidant is shown below in *Fig. 4*.

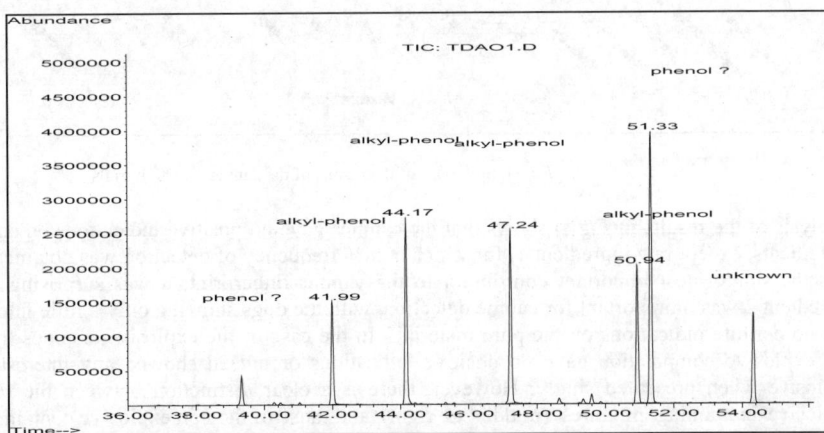

Fig. 4: GC chromatogram obtained from analysis of Anti-oxidant

Results of the canine analysis

Results of the canine olfaction trials are shown in *Fig. 5* below. The frequency of detection (%), averaged for both dogs is shown on the y axis against sample type on the x axis. The blue columns indicate samples on which the canines have given a positive indication, the red columns indicate samples in which the canine has shown interest, i.e. has not ignored but has not positively indicated. Non-explosive containing mixtures are grouped from the middle to the left, explosive containing mixtures from the middle to the right. The samples in the centre correspond to the pure explosives. The complexity of the mixtures generally increases from left to right, with raw ingredients on the left hand side and the full composition on the right. The results of the reward runs are shown in the column on the far right for comparison.

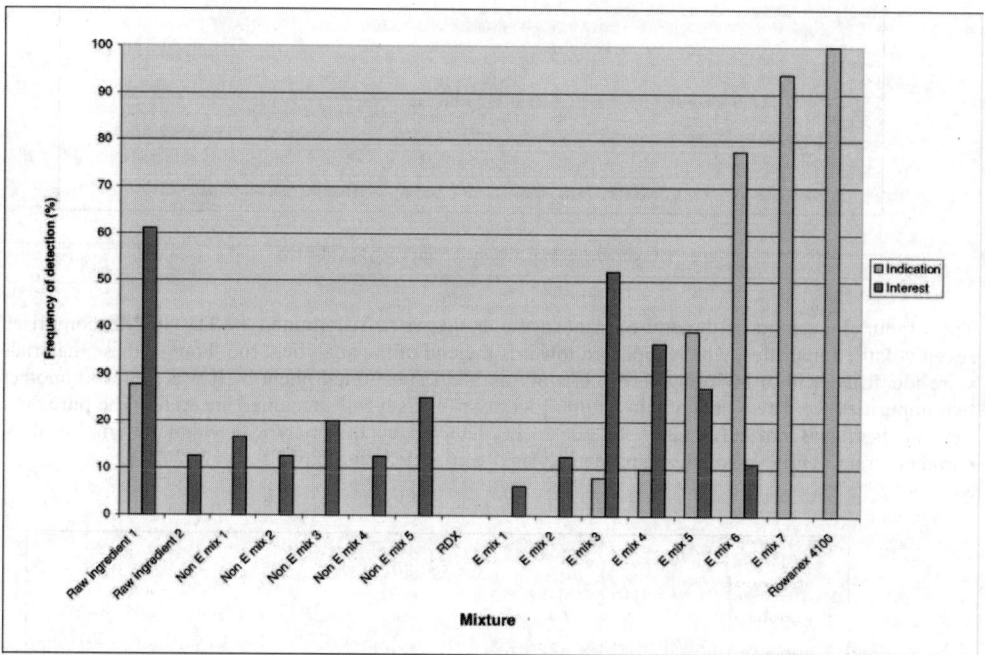

Fig. 5: Summary of the results of the canine olfaction trials.

Analysis of the results in *Fig. 5* shows that the canines gave no positive indications on any of the raw ingredients, except raw ingredient 1, for which a 28% frequency of detection was obtained. Given that it is the single most important contributor to the vapour fingerprint, it was surprising that the raw ingredient 2 was unimportant for canine detection, with the dogs showing only a little interest (12.5%) and no definite indications on the pure material. In the case of the explosive, canines trained on the full explosive composition gave no positive indications or indeed showed any interest in the pure explosive when presented singly. However, there is a clear distinction between the results of the explosive containing mixtures (middle to right) and those of the non explosive containing mixtures (middle to left). The canines recorded no positive indications and only relatively low levels of interest in any of the non-explosive containing mixtures whilst for the explosive containing mixtures, the frequency of detection increased as the composition approached that of the full explosive. High indication rates were only obtained for those mixtures with compositions matching closely that of the full explosive composition except for the omission of trace ingredients.

CONCLUSIONS

Results obtained in this study suggest that for canines trained on complex plastic explosives, the canine olfactory detection of a representative plastic explosive involves virtually the entire vapour fingerprint of this material. This would explain the high specificity of canine olfaction. Depleted mixtures lacking one or more ingredients failed to elicit significant levels of detection from canines used in the study. This was despite instrumental analysis showing these mixtures to be virtually identical to the complete mixture. Canines only began to reliably detect the depleted mixtures which were identical to the complete mixture except for the lack of one or more minor ingredients.

The inference from this is that the canine is a highly efficient pattern recognition detector, capable of detecting many compounds of varying chemical composition and volatility simultaneously. If trained on a highly complex mixture, it uses a high proportion of the numerous compounds present in the odour profile on which to make an identification, rather than relying on a sub section. This explains the high specificity of the canine in detecting plastic explosives. Although apart from the involatile explosive itself the majority of components in the headspace are not unique, the canine overcomes this by using a sufficiently high number such that the overall pattern is.

The presence of bulk RDX explosive, including its associated volatile impurities, are a necessity for the canine detection of the representative plastic explosive used in this study. This study indicates that the small suite of vapours associated with RDX, which comprise but a small subset of the entire vapour envelope of explosive, play an important role in the canine detection of this material. Canines trained on the full explosive failed to respond to pure RDX alone, despite the fact that RDX is the major component by mass. Further work is required however to determine whether or not the canine olfactory process involves detection of pure RDX vapour itself or solely the more volatile associated impurities.

Implication on the design of future explosive vapour detectors

The results of this study suggest that it may be feasible to mimic canines and effect the instrumental vapour detection of plastic explosives through the parallel detection of a number of volatile compounds emitted from the binder, of varying degrees of ubiquity. This challenging demand of the instrument is likely to require the use of pattern recognition techniques. This would utilise the entire vapour fingerprint or a small highly characteristic subset of volatile components, as opposed to detection solely of the explosive vapour itself

ACKNOWLEDGEMENTS

The assistance of Don Lavelle at Bae Systems Glascoed in preparing the explosive samples is gratefully acknolwedged, as is the contribution of Dr Marian Langford in the initial planning of this work.

REFERENCES

1.   'Analysis of Explosive Vapour Emission to Aid the Development of Vapour Detection Equipment', Proceedings of the 7th International Symposium on the Analysis and Detection of Explosives, Edinburgh, June 2001
2.   Steinfeld J., Wormhoudt J., Annu Rev, *Phys. Chem.*, **1998**, *49*, 203
3.   Villberg K., Veijanen A., *Anal. Chem.*, **2001**, *73*, 971

DEVELOPMENT AND CHARACTERIZATION OF EXPLOSIVE STANDARDS FOR TRAINING AND TESTING OF VAPOR-PHASE-DETECTORS

GERHARD HOLL

Bundeswehr Research Institute for Materials, Explosives, Fuels and Lubricants (WIWEB), Großes Cent, 53913 Swisttal, Germany

ABSTRACT

Several promising technologies for the detection of explosives are in development, each with its strengths and weaknesses.

This work reports on the application of using standard devices with explosives in order to produce user-defined authentic „odor signatures" to investigate the sensitivity and selectivity of different chemical sensors for explosive vapor detectors (EVD).

We achieved this aim by placing small quantities of explosives on the top of well defined inert metal surfaces (micro-mass-test samples – MPKs) without changing the chemical composition of the standard material. The emission rates of the explosive related chemicals (ERC) were evaluated at different temperatures and surfaces. With this data we were able to calculate the service life of this standards. Explosives of the MPKs have authentic chemical composition as explosives to be investigated but are non-hazardous when being tested in approved hazard classification tests. The chemical vapor signatures of different explosives have been verified by instruments and canine (K-9) detection.

INTRODUCTION

Explosive detector designs relevant to counterterrorism are based on a number of physical, chemical and mechanical properties.[i] This paper is focused on the evaluation of detectors based on vapor trace analysis.[ii] The systems detecting explosives by specific bulk properties (X-ray, neutron based systems, quadruple resonance or computing tomography) are not subject of this work.

43

M. Krausa and A. A. Reznev (eds.),
Vapour and Trace Detection of Explosives for Anti-Terrorism Purposes, 43-50.
© 2004 *Kluwer Academic Publishers. Printed in the Netherlands.*

The vapor phase detection of explosives is and will be a big challenge for two reasons:

1) The chemicals within a composition, which can be used to react in a short period of time to produce large amount of hot gases and/or blast waves – general definition of explosives –, do not have a specific chemical structure and functionality. That means different chemical sensors and sampling techniques are needed to analyze the explosive related chemicals (ERC) from the vapor phase.

2) The low vapor pressure and the high affinity of the high explosive molecules for different kinds of surfaces has been the reason why especially plastic based explosives including Semtex were considered as non-detectable.

The discovery of an „Explosive Threat" by vapor based explosives analytical systems from pure vapor state, explosive particles or carriers (dust specks) means to identify the explosive or the explosive related molecule (impurities and/or additives) within different chemical backgrounds *(Fig. 1)*.

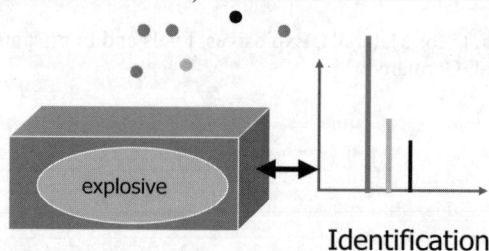

Identification

Fig.1: Identification of hidden explosives

The first step of a „sniffer" will sample the incoming air by drawing it over a surface onto which the sticky molecules attach themselves. In addition to the vapor pressure several other parameters will affect the possibility to catch a sufficient amount of characteristic molecules of ERC in the vicinity of an explosive: How far away we are from the equilibrium conditions (dilution, temperature etc.); what type of confinement was used (transportation mechanism); is there a detection gap caused by the affinity of the molecules to big surfaces; how will the sensitivity of the detector be influenced by interfering materials and impurities, etc.?

There are a lot of questions which should be analyzed for specific operational scenarios. In any case the challenge will be to match the detection strategy with the properties of the specific explosive compound present. To evaluate the wide variety of detection systems under realistic conditions it is necessary to have a mobile test equipment of explosives which can be easily used in many applications.

VAPOUR SIGNATURE OF EXPLOSIVES

Some detector systems, for example Ion Mobility Spectrometry (IMS), are capable to identify lowest quantities (< 1 ng) of chemical substances. With the location of explosives as well as objects with explosives at different places like buildings, vehicles, persons, above soil (mines), nevertheless one is dependent on efficient mobile well proved sensor techniques[iii]. Each individual sensor has to be trained (validated) for the different energetic formulation. Practically characteristic explosives as AN, EGDN, PETN and RDX are analysed. As a consequence explosives which do not contain these key molecules are not detectable by a usual equipment. To evaluate and to improve sensor technologies one has to handle different explosives under realistic operational scenarios.

For that reason the Bundeswehr Research Institute for Materials, Explosives, Fuels and Lubricants (WIWEB) has defined certain requirements that a test sample, especially with small quantities of explosives (micro-mass-test sample / MPK), should keep:

1. Through transportation of test samples, also in bigger quantities (at least 100 pieces) no dangers exist with respect to explosion, flash point, and virulence.
2. While handling the micro-mass-test samples the rules of the national Explosives Law will not apply.
3. Different kinds of explosives can be deposited safely on the surface.
4. Persons who work with the test sample do not come into contact with the explosives. There should be no possibility for cross-contamination.
5. The sample holder as well as the contaminated surface can become inert, i.e. without background smells of explosives.
6. Using exact geometry and defined quantities of explosive particles on a well defined surface, different source-strengths of evaporating substances of ERC are to be specified. The temperature dependence of the emission rate should be known.
7. The test samples can be hidden at different places. Their loss can be tolerated.

The most important requirements for explosive standards are summarized in *tab.1*.

Tab. 1: Requirements of explosive standards for vapor-phase detectors

- Authentic odor signatures as parent explosive
- Non-hazardous material (not Class 1 material)
- No handling of the material (substances are toxic)
- Different explosives can deposited safely
- Certification of the service life
- Avoid cross-contaminations with other explosives

The big advantage of using explosive standards as defined is to have the capability to evaluate the test procedure on site. For example if dogs are working in different environment to locate mines or unexploded ordnance (UXO) the standards can be hidden within the area. So one can easily examine whether the canine will still work properly or not (*Fig. 2*).

Fig. 2: How to use micro mass explosive standards for vapor detection

CHARACTERIZATION OF STANDARDS FOR THE VAPOR DETECTION OF EXPLOSIVES

The production of the micro-mass-test sample (MPK) prohibits the utilization of organic materials. To make the test samples inert, its individual components were heated up to 600 K over a time period of one to two hours. It had to be guaranteed that when handling the device a contact with the explosive is not possible. In *Fig. 3* one possible design of the micro-mass test sample is shown.

To put the explosive on the surface of the metal-husk, the material must be shredded mechanically with a suitable and save procedure. The nature of the surface as well as the size-distribution of the explosive particles determine the loading characteristics. Investigations with HPLC show that the quantity of explosive of each standard has a variation of less than 10 wt. %. The deposit procedure can be performed without using additional chemicals at ambient temperature.

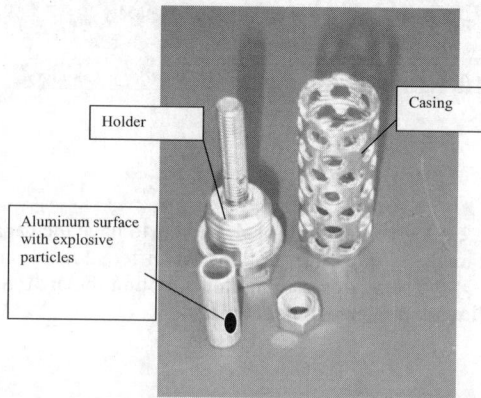

Fig. 3: Micro-mass-test sample; the explosive particles were stored on the surface of the aluminum shells

With the help thermogravimetric measurements (TGA) one can estimate the sublimation rate of solid TNT with a well known surface. At constant surface area the sublimation rate (S) is constant. Measurements at different temperatures enables us to calculate the temperature dependence of the sublimation rate witch follows the law of Clausius-Clapeyron

$$\ln S = \ln C - \frac{H}{RT} \qquad\qquad (1)$$

S	Sublimation rate	$[\mu g/(h*cm^2)]$
C	Constant	$[\mu g/(h*cm^2)]$
H	Sublimation enthalpy	$[kJ/mole]]$
R	Gas constant	$[kJ/(K*mole)]$
T	Temperature	$[K]$

From the least square evaluation of the experimental data we have computed the sublimation rate for different temperatures (*Tab.2*)

Tab.2: Sublimation rates as a function of temperature (ln C = 21,406 and H = 227 J/g.)

Temperature [°C]	Sublimation rate $[\mu g/(h*cm^2)]$
30	2,7
20	1,4
10	0,6

Especially at 10 °C the emission rate of TNT from the test sample is in the range of 0,5 ng/sec. This value can be influenced by the temperature and the size (surface) of the explosive particles. At 30 °C 50 percent of the initial mass of TNT evaporates after 50 hours (this is the service life of that TNT standard). Some characteristics of the micro-mass test sample (MPK) are shown in *Fig.4*.

- Loading 100...5000 µg /carrier
- Tested with 14 different solid explosives
- Emission rate of ERC, 293K (1)
  TNT: 0,3   ng/sec*cm²
- CL20:   0,0002 ng/sec*cm²
- Service life TNT (303 K, 1 mg):
  t(0,5) = 50 hours

Dr. Ticmanis, WIWEB, DTA-TG

•MPK in a sample holder for testing and training canine

Fig. 4: Some characteristic data of the micro-mass test sample (MPK)

APPLICATIONS OF MICRO-MASS TEST SAMPLES (MPK)

Electrochemical detection of explosives

The challenge still remains to develop some hand-held detector systems which is able to detect small quantities of explosives in operational scenarios. Electrochemical sensors may be a promising way for future control units. The setup of the electrochemical system which was investigated was developed at the Fraunhofer-Institut für Chemische Technologie (ICT)[iv].

To evaluate the characteristics of analytical tools, for example selectivity, sensitivity, response time etc., a gas-equipment has to be used which has the capability to generate different concentrations at low trace levels ($\leq$ 100 ppb). This is a time consuming and complicated process. With the help of explosive standards one can easily check a sensor system with different explosives with specified emission rates of volatile components (*Fig. 5*).

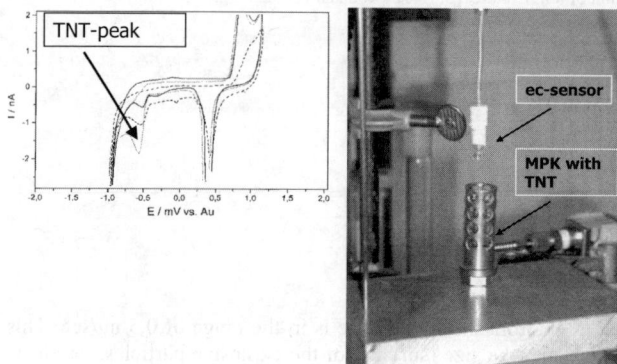

Fig. 5: Experimental setup (electrochemical sensor above the micro-mass test sample with TNT) and the cyclic voltammograms of TNT from vapor phase at room temperature as a function of time.

At approximately -0,5 V vs. Au TNT and other nitroaromatic compounds are reduced electrochemically. This signal was used to evaluate the electrochemical sensor characteristics.

Detecting explosives by means of sniffing dogs

In Europe dogs have been used for the detection of mines and explosives by military forces since the end of the second world war. Canine olfaction is an effective technique for explosive vapor detection. The major problem with explosive detecting dogs is that we neither understand in detail how they find the target nor we know the limits of their effectiveness; there is no scientific understanding up to now. So major research efforts are to be done to understand the mechanism by which canines achieve explosive detection.

For this reason we used MPKs with different kinds of explosives and registered the behavior of the dogs while working at the test facility (*Fig. 6*).

Fig. 6: Sniffing dog at work at a table with different sample holders for MPKs. The MPK could be loaded with explosive or not.

Nearly 100 teams took part in a round-robin test. One of the results of that investigation is illustrated in *tab.3*.

Tab. 3: Percentage of sniffing dogs with a correct indication of different explosives; the loading varies in µg range. TNT and DNT are recrystallized compounds.

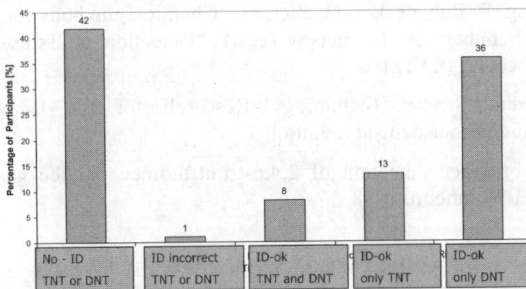

36 percent of sniffing dogs which were trained with commercially available TNT indicated DNT as an explosive. With additional experimental data we found out, that as far as we do not know the characteristic ERC with the highest response concerning the dog´s nose,

- the characteristic ERC with the highest response concerning the dog's nose,
- the authentic odor of the signal target odor (standards for training),
- the emission rate of the ERC (affect the detection threshold)
- and the type of explosive source (punctiform or diffuse)

play an important role when searching for explosives and chemical threats in vehicles, containers, trains, boats, etc. with sniffing dogs.

CONCLUSIONS

In the real world's environment the amount of different hazardous substances and their range of concentrations, which has to be controlled with optimized detection systems makes it clear how complex this mission is. Technical equipment or trained canines need chemical standards to enable a reliable identification of explosives. The application of micro-mass test samples (MPK) could be a good alternative for pure explosives. The evaluation and quality control of vapour based detectors can easily be executed while using in operational scenarios.

ACKNOWLEDGEMENTS

The financial support of the Bundesamt für Wehrtechnik und Beschaffung (BWB) is gratefully acknowledged. The author thanks the "Schule für das Diensthundewesen der Bundeswehr, Mr. Görgen" for his support to enforce the round robin test.

REFERENCES

1. U.S. Congress, Office of Technology Assessment, Technology against Terrorism: The Federal Effort, OTA-ISC-481 (Washington, DC:U.S. Government Printing Office, July 1991).

2. M. Krausa, H. Massong, P. Rabenecker, H. Ziegler, "Chemical methods for the detection of mines and explosives", in: H. Schubert, A. Kuznetsov (eds.), "Detection of Explosives and Landmines", Kluver Academic Publishers (2002) 1..19.

3. A Survey of Current Sensor Technology Research for the Detection of Landmines, http://diwww.epfl.ch/lami/detec/susdemsurvey.html.

4. M. Krausa, K. Schorb, Trace detection of 2,4,6-trinitrotoluene in the gaseous phase by cyclic voltammetry, Wiley-VCH Weinheim, 1998.

TRACKING THE TERRORISTS: IDENTIFICATION OF EXPLOSIVE RESIDUES IN POST-EXPLOSION DEBRIS BY LC/MS METHODS.

JEHUDA YINON, XIAOMING ZHAO AND ALEXEI GAPEEV

National Center for Forensic Science, University of Central Florida, Orlando, FL 32816, USA

ABSTRACT

In order to identify the type and origin of explosives in post-explosion debris, several groups of explosives, including TNT, RDX, nitroglycerin, PETN, EGDN, ammonium nitrate and some other inorganic oxidizers, were studied using LC/MS with electrospray and APCI ionization, and MS/MS for further confirmation of identified ions.

Characterization and origin identification of 2,4,6-trinitrotoluene (TNT) was done by means of its by-product isomer profile, including isomers of trinitrotoluene, dinitrotoluene, trinitrobenzene and dinitrobenzene, which are all by-products in the industrial production process of TNT.

Nitrate ester explosives (EGDN, NG and PETN) were studied by ESI- and APCI-LC/MS, using post-column additives. Explosives in mixtures could be identified at levels down to 5 ppb.

Formation processes of adduct ions in electrospray mass spectrometry (ESI-MS) of RDX were studied and origin of these adduct ions was determined.

A group of inorganic oxidizers were investigated. It was found that under certain conditions (temperature of heated capillary and selection of positive- vs. negative-ions) it is possible to identify these oxidizers by ESI-MS. These compounds produced cluster ions, which contained the entire oxidizer molecule.

INTRODUCTION

One of the tasks in the investigation of a terrorist bombing is to identify the type of explosive used in the bombing and if possible, its origin. This kind of information will assist the investigators and possibly lead to the apprehension of the suspects. The explosives used in terrorist bombings come from various sources, such as standard military explosives (stolen from military installations), commercial explosives used in mining and construction, explosives taken from landmines which were dug up, and improvised explosives, including

51

M. Krausa and A. A. Reznev (eds.),
Vapour and Trace Detection of Explosives for Anti-Terrorism Purposes, 51-62.
© 2004 *Kluwer Academic Publishers. Printed in the Netherlands.*

manufacture of explosives in clandestine laboratories. It is therefore of major importance to be able to identify and characterize the explosive residues, which in many cases are found in minute amounts. We have studied several groups of explosives, including 2,4,6-trinitrotoluene (TNT), 1,3,5-trinitro-1,3,5-triazacyclohexane (RDX), nitrate ester explosives, such as pentaerythritol tetranitrate (PETN), nitroglycerin (NG) and ethylene glycol dinitrate (EGDN), and inorganic oxidizers, such as ammonium nitrate. We have found that the analytical method of choice for identification and characterization of all studied explosive residues -from the point of view of sensitivity and selectivity- is LC/MS - both electrospray ionization (ESI) and atmospheric pressure chemical ionization APCI]) and tandem mass spectrometry (MS/MS). The instrument used was a Thermo-Finnigan LCQ-DUO LC/MS/MS system.

CHARACTERIZATION AND ORIGIN IDENTIFICATION OF TNT

A method, using LC/MS-APCI in the negative-ion mode, has been developed for characterization and origin identification of 2,4,6-trinitrotoluene (TNT) by means of its by-product isomer profile [1]. These isomers include isomers of trinitrotoluene (TNT), dinitrotoluene (DNT), trinitrobenzene (TNB) and dinitrobenzene (DNB), which are all by-products in the industrial production process of TNT.
HPLC separations were done with a Restek reversed-phase Allure C_{18} column (150x3.2 mm, 5 μm particle size), with mobile phases methanol-isopropanol-water (35:15:55) and methanol-water (50:50), at a flow rate of 0.4 mL/min and with sample injection volume of 10 μL.
Fig. 1 shows the mass chromatograms of a standard mixture of six TNT isomers and a mixture of three DNB, four DNT and one TNB isomers in methanol-water (50:50) at a concentration of 1 μg/mL. TNT samples from different and similar origins were analyzed in order to test their characteristic profile of by-products. Fig. 2-4 show the LC/MS-APCI mass chromatograms of TNT samples from Russia, China and from a landmine, respectively. Fig. 5 shows the LC/MS-APCI mass chromatograms of a 4-times crystallized TNT sample, which proves that purification of TNT can eliminate all the by-products. However, such purification is not being done by the manufacturers because of economic reasons and because the impurities have a positive effect on the casting properties of TNT.

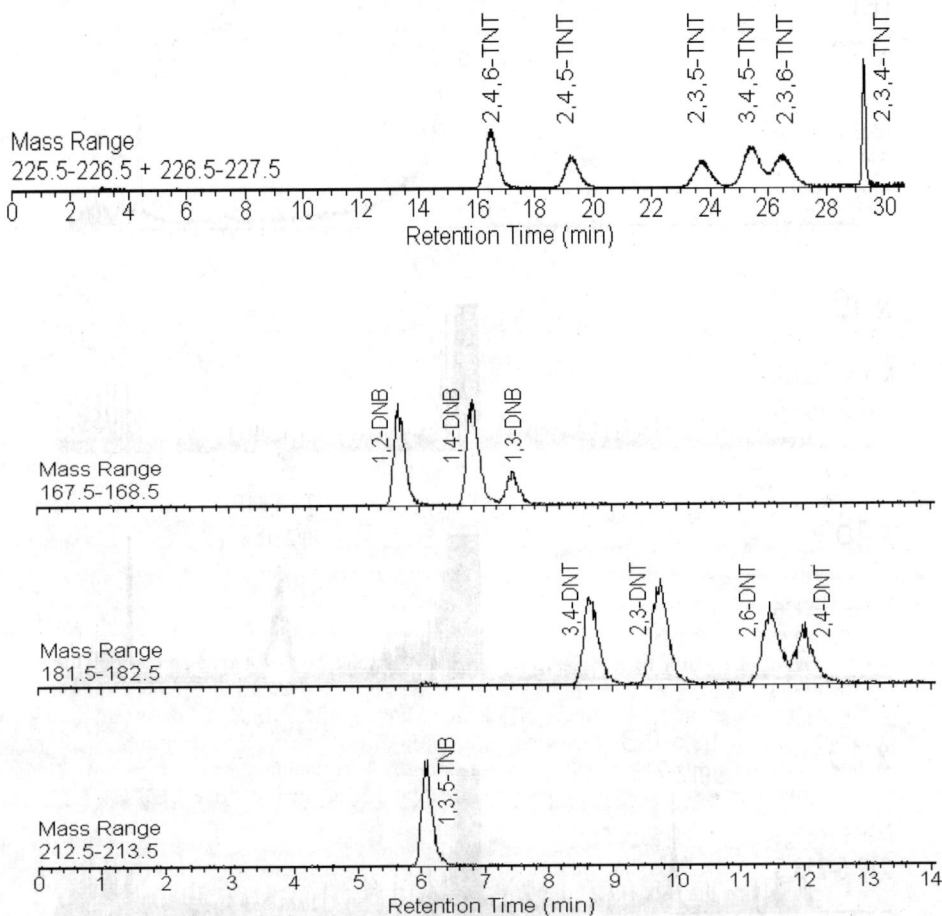

Fig. 1: Mass chromatograms of a standard mixture of six TNT isomers and a mixture of three DNB, four DNT and one TNB isomers in methanol-water (50:50) at a concentration of 1 µg/mL.

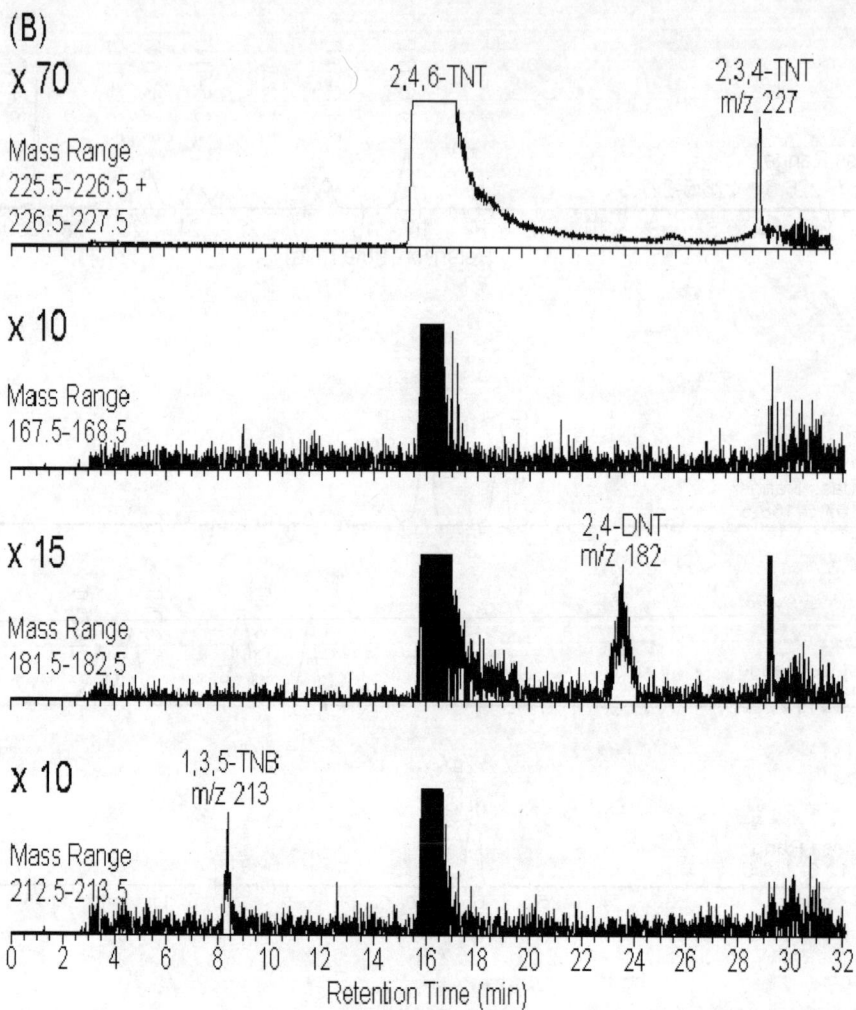

Fig. 2: LC/MS-APCI mass chromatograms of a TNT sample from Russia

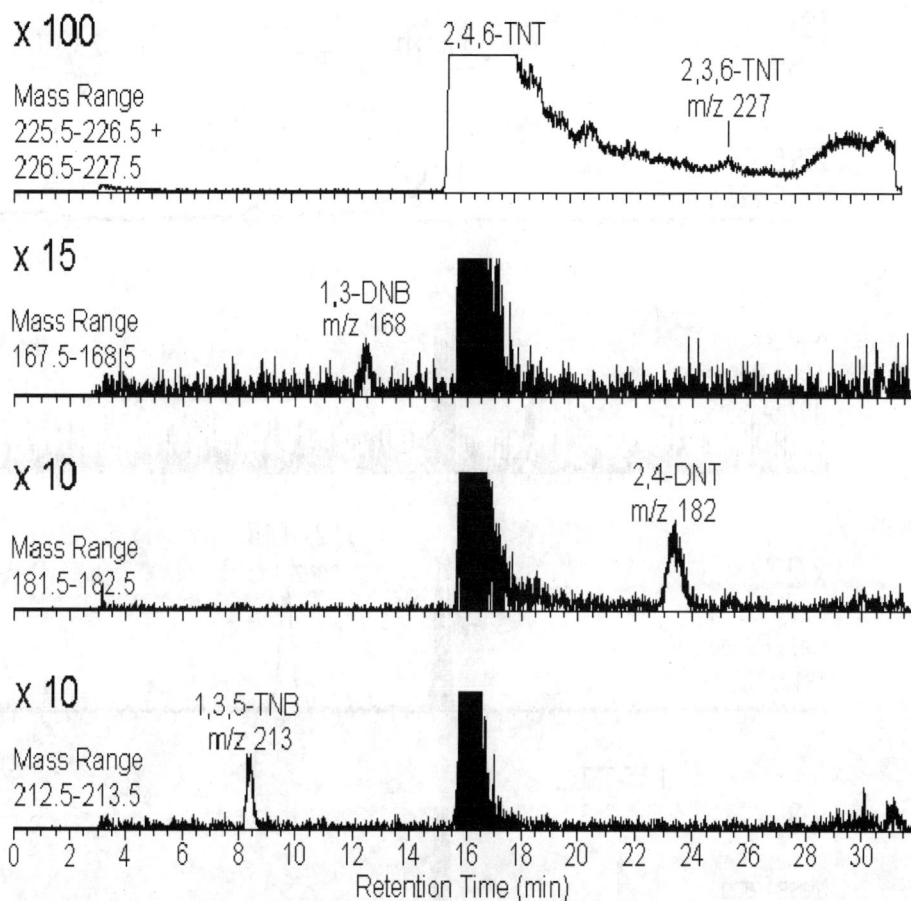

Fig. 3: LC/MS-APCI mass chromatograms of a TNT sample from China.

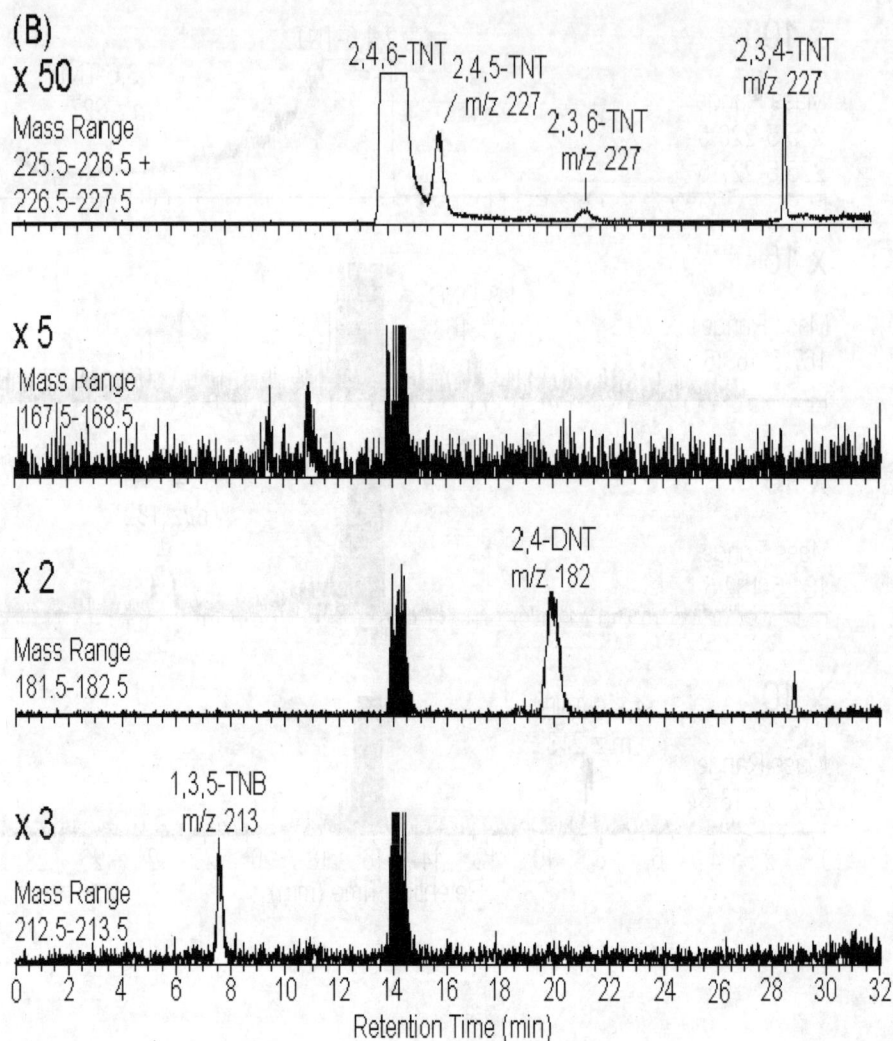

Fig. 4: LC/MS-APCI mass chromatograms of a TNT sample from a landmine.

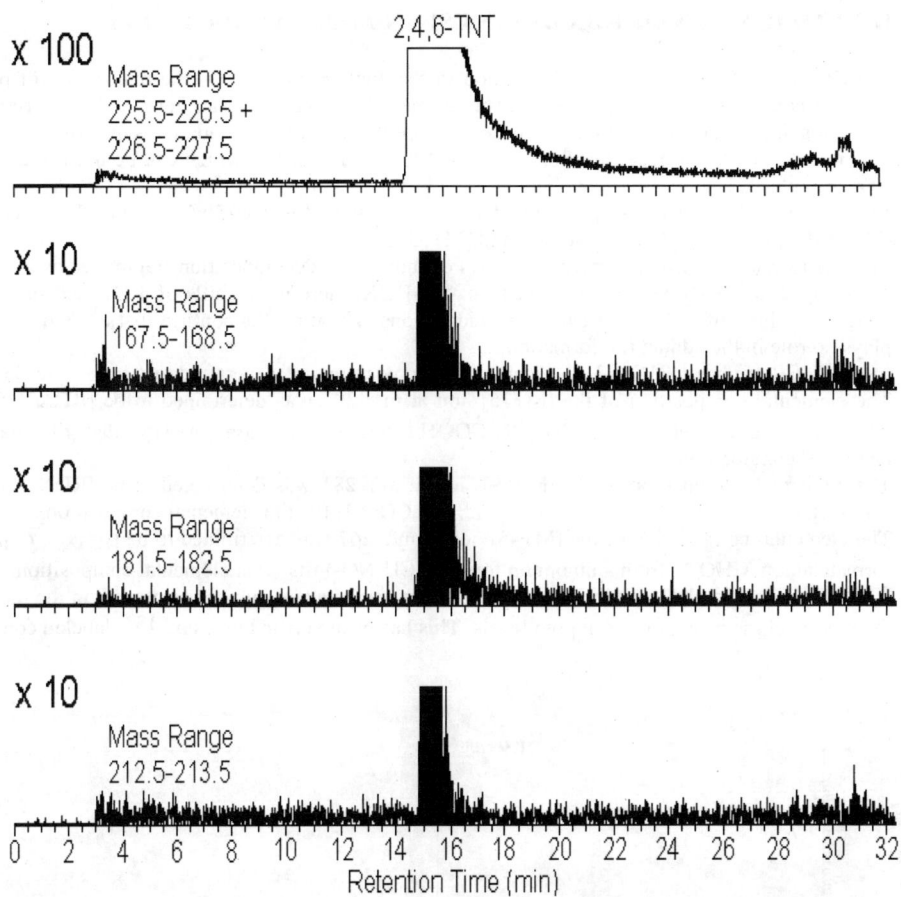

Fig. 5: LC/MS-APCI mass chromatograms of a 4-times crystallized TNT sample.

IDENTIFICATION OF RDX BY ITS ESI AND APCI ADDUCT IONS

In LC/MS of RDX, attachment of an anion to the analyte molecule is the major way of producing characteristic ions in ESI and APCI. The formation of RDX cluster ions and the origin of the clustering agents have been studied in order to understand the role of additives in formation of adduct ions in ESI and APCI mass spectra, and to find typical impurities which can be used to characterize RDX samples [2].

Fig. 6 shows the ESI mass spectrum of RDX in methanol-water (50:50). Typical adduct ions are [2M+75]⁻, [2M+59]⁻, [M+75]⁻, [M+59]⁻ and [M+45]⁻.

In order to establish whether the adduct ions contain RDX decomposition fragments, mass spectra of $^{13}C_3$-RDX and $^{15}N_6$-RDX were recorded. In $^{13}C_3$-RDX, there were shifts of 3 Da, and in $^{15}N_6$-RDX, there were shifts of 6 Da, for each of the adduct ions. These results confirm that RDX decomposition plays no role in the adduct ion formation.

In a FT-ICR study of RDX [3] the elemental compositions of several adduct ions were determined. The elemental composition of the [M+75]⁻ ion at m/z 297 was determined to be [RDX + $C_2H_3O_3$]⁻. Glycolic (hydroxyacetic) acid, $CH_2(OH)COOH$ forms an anion, [M-H]⁻, that fits the $C_2H_3O_3$ elemental composition.

The elemental composition of the [M+59]- ion at m/z 281 was determined to be [RDX + $C_2H_3O_2$]⁻. The acetate anion, $C_2H_3O_2^-$, (from acetic acid, CH_3COOH) fits the elemental composition.

The elemental composition of the [M+45]⁻ ion at m/z 267 was determined to be [RDX + CHO_2]⁻. The formate anion, CHO_2^-, (from ammonium formate, CH_5NO_2) fits that elemental composition.

RDX was found to cluster with formate, acetate, hydroxyacetate, which are present in the methanol of the mobile phase as impurities at ppm levels. This has been verified by using 13C-labeled compounds.

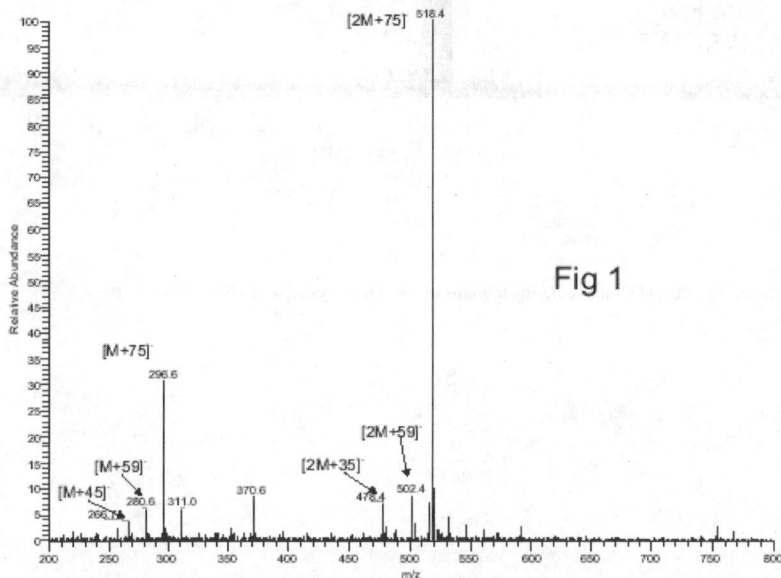

Fig. 6: ESI mass spectrum of RDX in methanol-water (50:50).

IDENTIFICATION OF NITRATE ESTER EXPLOSIVES

Nitrate ester explosives (PETN, NG and EGDN) were identified by ESI- and APCI-LC/MS, using post-column additives [4].

In ESI HPLC separation was obtained with a Restek Allure C_{18} column (100 x 2.1 mm, 5 μm particle size), using an isocratic mobile phase of methanol-water (70:30) at a flow rate of 0.15 mL/min. The additive, dissolved in methanol-water (70:30), was introduced post-column by a syringe pump (flow rate 5 μL/min) through a T union into the LC flow before entering the mass spectrometer.

Additives in ESI were ammonium nitrate, sodium nitrite, propionic acid and ammonium chloride, at respective concentrations of 0.005, 0.1, 0.2 and 0.1 mM. Respective adduct ions observed in the mass spectra were $[M+NO_3]^-$, $[M+NO_2]^-$, $[M+CH_3CH_2CO_2]^-$ and $[M+Cl]^-$.

In APCI HPLC separation was obtained with a Restec Allure C_{18} column (150 x 3.2 mm, 5 μm particle size), using an isocratic mobile phase of methanol-water (70:30) at a flow rate of 0.4 mL/min.

Additives in APCI were dichloromethane, chloroform, carbon tetrachloride and ammonium chloride, at respective concentrations of 0.2, 0.1, 0.05 % (v/v) and 0.3 mM. The adduct ion observed for all 4 additives was $[M+Cl]^-$.

Fig. 7 shows the LC/MS-APCI mass chromatograms of a mixture of 10 pg/μL PETN, 20 pg/μL NG and 2.5 ng/μL EGDN, with chloroform as additive.

Fig. 8 shows the LC/MS-ESI mass chromatograms of 25 pg/μL of a Semtex sample, with ammonium nitrate as additive.

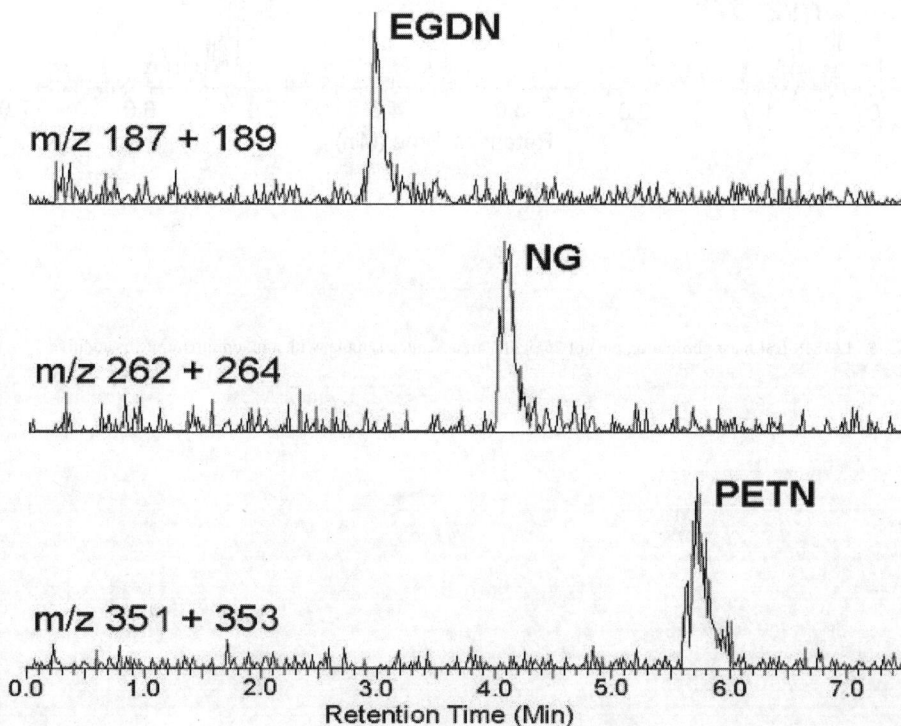

Fig. 7: LC/MS-APCI mass chromatograms of a mixture of 10 pg/μL PETN, 20 pg/μL NG and 2.5 ng/μL EGDN, with chloroform as additive.

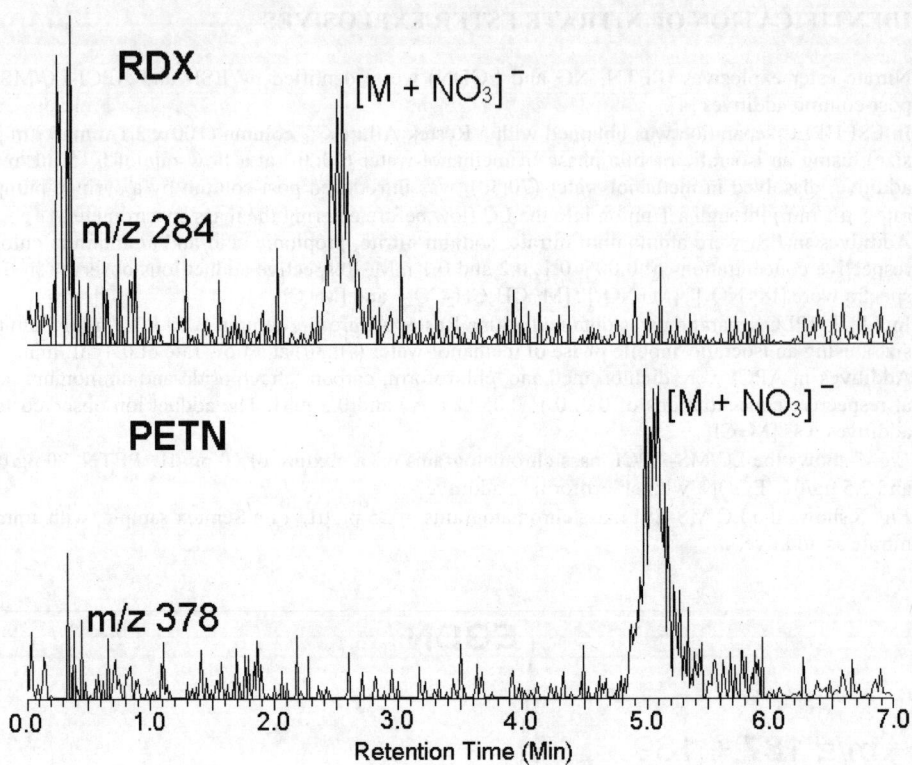

Fig. 8: LC/MS-ESI mass chromatograms of 25 pg/µL of a Semtex sample, with ammonium nitrate as additive.

CHARACTERIZATION OF AMMONIUM NITRATE AND OTHER INORGANIC OXIDIZERS

Inorganic salts have been used extensively as oxidizing agents in various explosive mixtures, such as ANFO, dynamites, etc. We have studied ammonium nitrate and an additional group of inorganic oxidizers and found that under certain conditions (temperature of heated capillary and selection of positive- vs. negative-ions) it is possible to identify these oxidizers by ESI-MS [5, 6]. These compounds produced cluster ions, which contained the entire oxidizer molecule. Isotopically labeled compounds and MS/MS were used to confirm the suggested ionic compositions.

Fig. 9 shows the positive-ion ESI mass spectra, at a temperature of 100° C of the heated capillary, of 1 mM each of NH_4NO_3, $NH_4{}^{15}NO_3$, ${}^{15}NH_4NO_3$ and $NH_4N{}^{18}O_3$ in methanol-water (50:50).

Fig. 10 shows the positive and negative-ion ESI mass spectra of 1 mM sodium perchlorate in methanol-water (50:50). Temperature of the heated capillary 220° C.

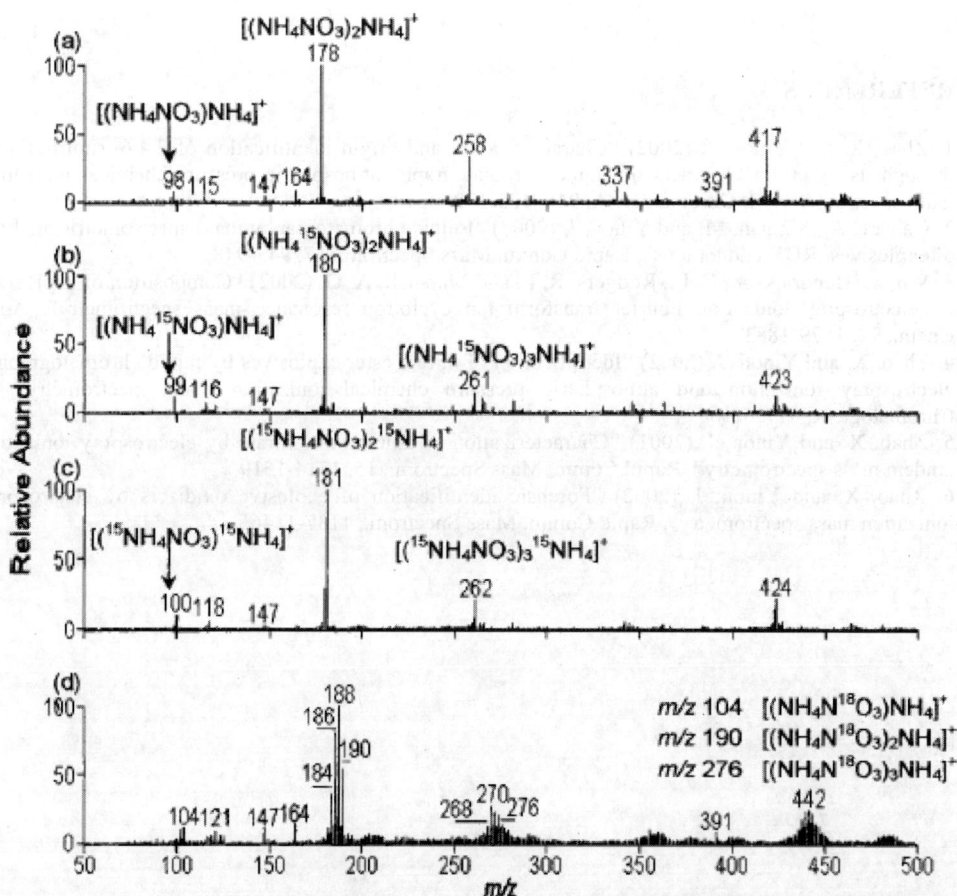

Fig. 9: Positive-ion ESI mass spectra, at a temperature of 100° C of the heated capillary, of 1 mM each of NH_4NO_3, $NH_4{}^{15}NO_3$, ${}^{15}NH_4NO_3$ and $NH_4N{}^{18}O_3$ in methanol-water (50:50).

CONCLUSIONS

a. TNT can be characterized by its profile of isomers of TNT, DNT, DNB and TNB in LC/MS-APCI.
b. RDX can be identified by LC/MS-ESI, using ammonium nitrate as post-column additive.
c. Nitrate ester explosives can be identified by LC/MS-ESI, using ammonium nitrate, sodium nitrite or ammonium chloride as post-column additives, or by LC/MS-APCI using methylene chloride, chloroform, carbon tetrachloride or ammonium chloride. Identification is confirmed by MS/MS.
d. Ammonium nitrate and other inorganic oxidizers can be identified by a series of cluster ions in their negative-and positive-ion ESI mass spectra at various temperatures of the heated capillary.

REFERENCES

1. Zhao, X. and Yinon, J. (2002) "Characterization and origin identification of 2,4,6-trinitrotoluene through its by-product isomers by liquid chromatography-atmospheric pressure chemical ionization mass spectrometry", J. Chromatogr. A, 946, 125-132.
2. Gapeev, A., Sigman, M. and Yinon, J. (2003) "Liquid chromatography/mass spectrometric analysis of explosives: RDX adduct ions", Rapid Comm. Mass Spectrom., 17, 943-948.
3. Wu, Z., Hendrickson, C. L., Rodgers, R. P. and Marshall, A. G. (2002) "Composition of explosives by electrospray ionization Fourier transform ion cyclotron resonance mass spectrometry", Anal. Chem., 74, 1879-1883.
4. Zhao, X. and Yinon, J. (2002) "Identification of nitrate ester explosives by liquid chromatography-electrospray ionization and atmospheric pressure chemical ionization mass spectrometry", J. Chromatogr. A, 977, 59-68.
5. Zhao, X. and Yinon, J. (2001) "Characterization of ammonium nitrate by electrospray ionization tandem mass spectrometry", Rapid Comm. Mass Spectrom., 15, 1514-1519.
6. Zhao, X. and Yinon, J. (2002) "Forensic identification of explosive oxidizers by electrospray ionization mass spectrometry", Rapid Comm. Mass Spectrom., 1137-1146.

REMOTE EXPLOSIVE SCENT TRACING

– a method for detection of explosive and chemical substances.

RUNE FJELLANGER, CAND. SCIENT.

NOKSH AS, Postboks 57, 5201 Os, Norway

INTRODUCTION

Remote Explosive Scent Tracing (REST) involves the concept of transferring a target odour to an animal detector (dogs), using a specially designed filter. The filters are transferred to dogs trained to identify specific target odours such as drugs or explosives. The contaminated filters are produced/sampled at the suspect site by vacuuming air through them. Application for the detection of landmines involves vacuuming an area of land suspected to contain mines, and requires that the dogs are trained to detect very low concentrations of target odour. Testing of the filters by the detector is undertaken in a laboratory environment, and involves a number of internal controls to ensure reliability.

The REST method as a technique for detection purposes was developed by the South African government, in the hands of Mechem Consultants, a division of Denel, (Pty), Ltd. at the end of the 1980s. The method was first referred to as MEDDS (Mechem Explosives and Drug Detection System) [4]. It has subsequently also been called EVD (Explosives Vapour Detection) by Norwegian Peoples Aid (NPA), Angola. The primary purpose of this method was to detect drugs, explosives and weapons at border crossing checkpoints. The system was utilized at the many checkpoints in South Africa at the borders with Swaziland and Lesotho, but also at the main border crossings to Mozambique, Zimbabwe and Botswana. Experience has shown that the system had the capacity to check a significant number of vehicles, train carriages and other cargo in a short time and required little personnel.

The target substances from e.g. explosives (mines) is collected by suction of air and dust particles from the ground surface, using a portable vacuum pump with filters placed either in a vehicle or carried as a backpack by a person walking. The filter cartridges shall be held close to the ground during scent trapping to ensure the collection of maximum contamination. When air is sucked close to the ground, dust and soil particles will also come into the filters. The concentration of strategic scent substances attached to soil particles can be 1 million times higher than in the air directly over the ground [6]. The filters are mainly made of coiled polyvinyl chloride (PVC) netting.

63

M. Krausa and A. A. Reznev (eds.),
Vapour and Trace Detection of Explosives for Anti-Terrorism Purposes, 63-68.
© *2004 Kluwer Academic Publishers. Printed in the Netherlands.*

The air/filter samples are typically collected over sections of 100 meters when the REST technique is applied to mine detection. The operator follows in the wheel tracks of mine-proof vehicles and collects samples from the ground area in front and on each side, within the reach of the sampling mechanism. Each filter placed in the collection box is marked with a detailed log describing position (GPS), date and time. Additional information includes temperature, humidity, wind strength and direction for each change of filter. The filters are subsequently sent for analysis by specially trained sniffer dogs. These will indicate whether target substances from mines/UXO exist, and relate these to geographic positions described by the GPS logging.

When REST was originally conceived and used operationally for mine detection in South Africa, through the early to mid-1990s, there was little documentation available of the original research on its development. Despite its apparent success as a very efficient technology for area reduction, it was never embraced by the mine clearance industry.

In the year 2000, the Geneva International Center for Humanitarian Demining (GICHD) initiated a broadly-based program of research on mine detection animals, which also includes attempts to further develop and understand the potential of REST technology. The Norwegian company, Norsk Kompetansesenter for Spesialsøkshund AS (NOKSH AS) was engaged for this purpose.

REST SYSTEM IN BOSNIA

In mid 2000, the GICHD also signed a contract with NOKSH AS regarding the conduct of three projects under the GICHD MDD study. One of the projects for GICHD addressed training methodology for REST. This comprised a series of activities and one was experimental training of 4 REST dogs. The project was completed in mid 2001 and the final reports were submitted to the GICHD in late 2001 [3]. In 2002, the research program was continued by practical experiments testing the REST method in Bosnia – Herzegovina.

The Balkan region was chosen as a particularly interesting area for several reasons; firstly, the REST system is not currently used in Europe. A successful REST programme exposed in Bosnia may draw greater attention to the system and result in a more widespread use of it among demining organisations. Secondly, the use of REST may have a very positive impact on resolving the mine problem in the Balkans. Thirdly, a REST capacity in Europe will inevitably be of good support to further research REST system improvements. It will facilitate a closer co-operation between the REST programme, the GICHD and the research organisations. Preliminary investigations in Bosnia suggest that both BiHMAC and NPA are very interested in developing a REST capacity in Bosnia, one that could potentially support the whole Balkan region.

Based on this it was decided to set up a pilot study designed to determine whether REST technology can be used for effective area reduction in Bosnia. Due to cool temperatures, heavy soils and wet summers, it is believed by mine clearance operators that mine detection by dogs is relatively difficult in Bosnia. As REST for mine detection depends on the availability of explosive vapours in the minefields, it seems likely that similar difficulties will apply to the use of REST in Bosnia. The pilot study also intends to investigate a number of related factors that potentially influence the ability to detect mines with filters and dogs.

METHODS

As a part of the study set up by NOKSH AS for GICHD, four dogs were put through a training program for gathering data and to get empirical evidence about the method. The following principles were implemented: i) to minimize the dependency on the handler,
ii) encourage independence in the dogs' search, iii) build extended search by progressively reducing the frequency of positive filters, iv) minimize anticipation of reward due to positive indication event [1], [2] and v) systematically reduce the detection threshold for TNT.

All training involved positive, reward-based shaping, using "clicker" training, a concept that allows very precise timing links to be made between actions of the dog and rewards provided by the trainer [7]. As training progressed, the clicker was used only to provide an instant of sound (application of principles i and ii above). Once the clicker was established as a conditioned reinforcement, ongoing reinforcement of the clicker was given intermittently using a variety of reinforcing stimulus (food, ball, praise, etc).

The training apparatus was a circular stand with 12 stainless steel arms, effectively a multiple-choice apparatus [5]. Fittings for boxes or filter cartridges were placed on the end of each arm. The stand could be revolved after search by a dog, allowing rapid adjustment of filter position. Filters could quickly be exchanged, or transferred between arms. The stand was regularly cleaned with alcohol, and all handling of equipment and filters was done using disposable plastic gloves. The stand was placed in a small room (4 x 4 m) with two entrance doors, a blind for trainers to hide behind, and a one-way window. Dogs were trained to enter from one of the doors, make one circuit of the stand sniffing at each filter, and exit the room from the same door. For any search, zero, one or two trainers could be in the room, either hidden behind the blind or present in the room. For the dogs, the only constants on any search event were the presence of the circular stand and the room itself. Dogs search the stand with no support or assistance from the trainers, except for the click reward when a correct positive indication is given. If the dog correctly determines that there were no positives on the current trial (i.e. it gave no indication), the dog was rewarded once it was outside the door. The operational requirement to have dogs indicate reliably without being rewarded was progressively introduced from an early stage in the training program.

The study used test minefields previously established by Norwegian Peoples Aid near Sarajevo and Mostar. Thus all mines used had been in the ground for long periods (up to several years).

Factors to be studied were:
- When the sampling machine encountered the mine (start, middle, end of the 60 sec sampling period)
- Total time filter was held over the mine (pass only, 1 sec, 2 sec, 5 sec)
- Weather variation (recorded at the time of sampling)
- Type of mine sampled (three)

Factors to be held constant were:
- Total sampling time (60 secs)
- Sampling vacuum rate (60 litres/sec passing through the filter)
- Depth at which mine was laid (within 10 cm of surface)
- Equipment (small petrol-driven vacuum machine, and the Mechem filter)
- Sampling procedure (operator walks slowly passing the vacuum nozzle back and forth across the ground)
- Number of mines sampled onto one filter (one)
- Testing (all filters tested with 4 dogs)

The sampling procedure involved the operator using a mine on the edge of the test field and walking either towards or away from the mine, in order to encounter it at the required sector of the sampling interval (beginning, middle, end). Weather factors recorded at the time of sampling were temperature and humidity at breast height. The temperature gauge was not shaded. All sampling was done in light winds or calm conditions, and at least 24 hours after heavy rain.

Two measures for probability of detection were available: i) the proportion of "positive" filters that were detected, and ii) the proportion of dogs that detected each "positive" filter (each dog was given two passes on the filter before it was declared as a negative). Every trial was performed blind (randomized), because no information on the sampling identity of the filters was provided to the personal conducting the tests until all testing was completed.

It was assumed that all of the four dogs were working at equivalent detection sensitivity and capability. Internal checks using known positives and negatives tested for reliability. A "miss" is a filter that is supposed to be positive, but which is not indicated by the dog after two passes. A "false positive" is a filter that is supposed to be negative, but is indicated by the dog as positive.

RESULTS

Overall, the proportion of mines found on 88 filters was 68%. More mines were missed in Sarajevo (where temperatures were lower and rainfalls stronger) than in Mostar. Almost all mines were found in the Mostar area.

No significant effects were found for
- position in the 60 sec sampling period (beginning, middle, end),
- whether the sampling nozzle passed over, or paused over, the mine (pass, 1, 2, 5 sec)
- type of mine (PMA3, TMA4 and TMM1).

Significant effects were detected with regards humidity (detection success was higher when humidity was lower) and temperature (detection success was higher at higher temperatures). The effect of humidity was stronger than the effect of temperature, and the two effects were not accumulative.

DISCUSSION

Although this was a pilot study, we conclude that the REST concept can be efficiently used to detect mines in Bosnia. The lowest proportion of mines found was in Sarajevo, and this is mainly due to the weather conditions and soil that contains a lot of clay. Experience from using ordinary mine detection dogs in Sarajevo has shown that it can be relatively difficult in some periods, especially in the autumn when it has been raining for some amount of time.

Clearly, there is a need for further development, or fine-tuning of the sampling process, which will particularly require giving attention to the effects of humidity. The effects of temperature may be less important, although operational field experiences combined with evidence from this study suggest that there is a level of minimum temperature below which sampling should not occur. For the moment, that temperature is considered to be 15°C.

Further tuning of the detection process is also clearly required, as the detection success was considerably lower than is necessary for operational use of REST technology. In this study, 4 dogs were used, and the probability of detection by those dogs varied. For the moment, the cause of that variation is not understood and requires further research. Certainly, an important requirement is to "tune" the dogs on prepared filters from the operational sites before detection begins. There was only limited opportunity to undertake such tuning for this study. Experiences from the few organizations training REST detectors suggest that training this type of detection dogs takes up to 12 months before the dogs are operational. However, it is also clear that considerable work needs to be invested in tuning the detector in order to obtain high levels of detection success. At the end of training, the dogs used in this study were giving detection success of about 95% [2], indicating that the success rate in this study could be considerably improved. Further research on how to fine-tune the detection process is clearly required.

New perspectives on REST: An effective tool for anti-terrorism.
In the year 2003, NOKSH AS is also playing a key role in a similar test program in Angola. These programs have become valuable sources of information on how physical parameters (temperature, humidity, etc.) influence vapour detection when sampling is conducted in an open-air environment.

While working with the REST technology, NOKSH AS has gathered insight into how the method can be applied to other and new areas as a detection system e.g. cancer diagnoses, oil leaks in power cables. In combination with an effective, quick and reliable training technique for producing an animal detector (dogs), this technology, vapour detection of explosives, can be transferred and used directly in other practical search areas.
In close cooperation with chemists and technical staff, NOKSH AS can supply the system with other selective filters (e.g. chemicals and other substances), and subsequently make the REST system into a prominent detection method for new departments which also includes anti-terrorism purposes.

REST has become the most reliable and cost-effective method in humanitarian demining industry searching for mines in mine suspected areas. The results of the training also indicate that it is the most time saving method. The REST method has demonstrated the capacity to search large areas with a minimum amount of time, while detecting substances at extremely low concentrations. The REST method alone or as a supplement to other techniques could represent a huge step forward when searching for explosives, weapons, ammunition and chemical threats in vehicles, containers, trains, boats, etc. It can also be used to check or verify if people have been in contact with such substances. Such tests can also be carried out without the objectives knowing that they are being tested, and more importantly, not knowing the results of such tests.

Acknowledgments

I would like to thank Espen Kruger Andersen, Fjellanger Dog Training Academy and Andalosie Sanjala, NPA Lubango both of whom assisted in the detection laboratory. Sampling was undertaken in collaboration with Norwegian Peoples Aid, Bosnia. I would also like to thank Elmir Tozo, Edin Avdic and Terje Berntsen for their support. Thanks also to Per Jostein Matre, Tor Helge Eiken (NOKSH AS) and Ian McLean (GICHD) for assisting in writing the text. The Bosnia project was funded as part of the broader GICHD dog research program.

REFERENCES

[1] CATANIA, C. (1992) Learning. New York: Prentice Hall

[2] FJELLANGER, R., (2001). Remote Explosive Scent Tracing. In: *Mine Detection Dogs Study,* GICHD, Geneva.

[3] FJELLANGER, R., (2003). General Principles of the Psychology of Learning. In: GICHD (Ed.) *Mine Detection Dogs: training, operations and odour detection.* GICHD: Geneva.

[4] JOYNT, V., (2003). Mechem Explosive and Drug Detection System (MEDDS). In: GICHD (Ed.) *Mine Detection Dogs: training, operations and odour detection.* Geneva: GICHD.

[5] KING, J. E., R. F. BECKER and J. E. MARKEE (1964). Studies on olfactory discrimination in dogs: (3) Ability to detect human odour trace. British Journal of Animal Behaviour 12(2-3): 311-315.

[6] PHELAN, J. & BARNETT, J.L. (2002). Chemical sensing thresholds for mine detection dogs. In: ORLANDO, F.L., DUBEY, A.C., HARVEY, J.F. & BROACH, J.T. (Eds) *Proceedings of the SPIE 16th Annual International Symposium on Aerospace/Defense Sensing, Simulation and Controls, Detection and Remediation Technologies for Mines and Minelike Targets,* 7, 532-543.

[7] PRYOR, K., (1999). *Don't shoot the dog!.* New York: Bantam Books.

DETECTION OF TRACES OF EXPLOSIVES BY MEANS OF SNIFFING DOGS

MIROSLAV STANCL

Explosia a.s., Research Institute of Industrial Chemistry, 53217 Pardubice – Semtin
Czech Republic

ABSTRACT

The contribution is focused on our experience with detection of hidden explosives, marked explosives and marking taggant by means of sniffing dogs. The capabilities of sniffing dogs were tested under different conditions to find hidden samples of explosives, releasing traces of vapours or particles. 11 samples of basic types of explosives, including DMNB taggant, were sought out by sniffing dogs within the framework of five examination scenarios.
The results obtained enable to make certain conclusions concerning the work of dogs in the field of detection of explosives.

INTRODUCTION

For years dogs have been successfully used to detect hidden explosives under various environmental conditions. Their main advantages are mobility, significant improvement over fixed detector installations and cheaper dog training program when compared with electronic detectors price. A dog is a special vapour detector because dog's powerful sense of smell, known as the olfactory ability, is about 30 – 40 times better than the said ability of human due to the size of olfactory epithelium surface. At dogs, this surface is covered with olfactory cilia attached to the receptors. The cilia float in a mucus that bathes the olfactory epithelium, so that the receptors are in direct contact with the air when they lie on the surface of the film. Odours dissolve in the mucus and trigger a transduction system within the cilia, which results in generation of an electrical signal within the receptor, which in turn passes to the olfactory bulb. The signals are processed there before being sent to the brain [1].
It was found that for actual vapour of the explosive or marking taggant, detection within ppb to ppt by properly trained dogs is possible [2].

M. Krausa and A. A. Reznev (eds.),
Vapour and Trace Detection of Explosives for Anti-Terrorism Purposes, 69-77.
© 2004 Kluwer Academic Publishers. Printed in the Netherlands.

BACKGROUND

Still more significant effort has been spent in the world within the last years on unification and establishment of widely accepted training and testing methodologies or protocols and minimum discretionary ability standards for explosives detecting dogs. In many countries the research in this field is focused on breeding, training, testing and evaluation of skills and certification of explosive detection dogs.

In 1990 the Bureau of ATF started a joint program with the U.S. Department of State and the Connecticut State Police to produce more effective explosives detection dog (K9).

ATF Canine Explosives Detection Program identified three major goals:
1) Systematic selection of explosives for training
2) Standardisation of testing and evaluation procedures
3) Elimination of cross-contamination of explosive odours

For training purposes, explosives were classified as nitro compounds (aliphatic and aromatic), nitrate esters and nitramines. For training of dogs scent box is used at first, and later the training wheel. The device consists of four containers on a rotating wheel. One can is positive sample container and the others contain no samples or samples with distracting odours (distractors). Training is considered successful when the dog ignores a food distractor in favour of the explosive odour stimulus. It is very important to avoid so called cross- contamination of one explosives sample with the odour of another. The use of standardised protocols and testing procedures provides accurate and credible results, and training methodology using food reward proved to be a rapid and effective means for dog training. [3].

After completing the training program, the dog/handler teams were used to test the lower limit of the dogs nitromethane vapour detection response. There were 503 tests of nitromethane and diluted nitromethane samples performed by means of standard ATF test wheel. It was found that the probability of detecting nitromethane remained high until the vapour pressure fell below ~ 1×10^{-6} microns. This would place detection limits in the part-per-trillion range (i.e. ~ 10^{-12}) [4].

At the University of Toronto, Canada, small animals were trained to detect C-4, Deta Sheet and Semtex by means of automated vapour injection system following pre-training with a passive vapour exposure device. Half the animals were trained using the vapour injection system with a preconcentrator in the target flow, while the other animals received no preconcentration. The results showed that the use of the preconcentrator lowered significantly false alarms and may have also a positive effect on correct detection of explosives vapour [5].

Russian specialists certified the quality of a large group of dogs trained for explosives detection. Qualitative evaluations were based on two criteria - indication of explosives samples or missing the blank samples on the search strip (a strip of open space of 5 m width and 70 – 80 m length). The explosives sample plant was a glass vessel of 10 ml volume, containing 1 g of TNT which was buried 15 cm deep into the soil. The analysis carried out allows to make two conclusions. Only the dogs making no misses of explosives plants (or max. 1 miss) may be used for explosives search in real conditions. The main part of the dogs taken for service without special testing must not be used for the search of explosives. It appears that the number of dogs suitable for similar search service is not more than 20 % [6].

French specialists conducted testing at airports in two areas – detection of explosives vapours and detection of traces of explosives. A protocol for detection of explosives vapours was developed which takes up the protocol by the ICAO (ICAO suitcase).

For the purpose of detection of traces of explosives the surface of ICAO suitcase was contaminated by means of operator's "dirty" gloves (explosives used: C-4, plastrite, formex, semtex and tolite).The tests all together showed average detection rates of 72 % and false detection rates of 5 %. For vapour detection the probability of detection (POD) is 60 % and for traces detection POD is 80 %. The following equation has been formulated:

$$POD = \frac{f(T, H)}{P}$$ where

POD ...probability of detection
T .... temperature (°C)
H ...... relative humidity
P ...... conditions at the location (such as dust)
[7]

In 1997, the DGCA (France) performed tests aimed at comparing the performance of dog handling teams and explosives detector Barringer Ionscan 400. The tests were conducted in two ways. 50 real bags guaranteed to be free from any explosives were prepared as "clean" bags and "dirty" bags.
In case of "clean" bags two experts prepared sample (the first of them placing explosive charge into the sachet and the sachet into the suitcase, and the second one moving the suitcase into the bag). The third expert handled the particle detector.
"Dirty" bags were prepared by one expert only. The results of tests are shown in the *Tab.1.*

Tab.1:

	"Clean" bags		"Dirty" bags	
	Dogs	Ionscan	Dogs	Ionscan
Detection rate	80 %	0 %	76 %	62 %
False alarm rate	11 %	0 %	38 %	14 %

The results show very good performance on the part of dog handling teams as compared to the electronic detector. This performance is comparable with the detectors available on the market that are very expensive. Effectiveness of a dog handling team is not absolutely constant because dog's behaviour depends on it's psychological state and environmental conditions. In addition of being an interim solution, the dog handling team detection method is a good complement to x-ray detection methods. It can be aptly applied to analyse quickly a batch of suspect bags [8].

A review of research results in the field of use of dogs for demining purposes is given in [9]. The information used come mostly from Auburn University, the US Army Corps of Engineers and Sandia National Laboratories. The research was concentrated on identification of substances from landmines that are detectable to dogs. TNT is one of the most common explosives in landmines (about 80 % of all), however TNT contains by products, it means impurities and degradation products such as dinitrobenzene (DNB) and dinitrotoluene (DNT). 2,4 DNT gave the most consistent response across all dogs studied, following by 1,3-DNB. The results were that the sensitivity of the dogs for odour with 2,4-DNT declined between about 200 and 1000 parts per-trillion (ppt). (100 ppt approximately equals 1 nanogram per litre). These sensitivity estimates approach the limits of current technology.

Other important results:

- Dogs detect relatively very low concentrations of target odour in the presence of strong masking odours. At a target odour concentration of about 1 ppb, concentration of masking odour needed to be about 20 ppm before detection performance began to be affected.
- The time delay (or forgetting) in the ability to retain skills without refreshing training ranges between 14 and 120 days.
- Ten odours were the maximum tested in the experiment and it seems that some or even many more odours could be trained.
- Dogs can work effectively for at least 90 – 120 minutes of continuous searching, however temperature affected dog's performance – high temperatures can decrease dog's detection capability.
- The substance, the dogs respond to is 2,4-DNT, not TNT as such. DNT seems to be important in the so-called "detection signature" for a landmine.
- Detection ability of dogs remains better than detectors by several or even by many orders of magnitude.

EXPERIMENTAL WORK

Place and methodologies of tests performance

The tests were conducted on May 23, 2002 in the premises of the Police of the Czech Republic, Police Presidium of the Czech Republic, Section of Service Cynology and Hippology in the training centre of dog handlers and service dogs in Býchory in cooperation with the management of the centre of service dog handlers and workers of the Research Institute of Industrial Chemistry (VÚPCH). Special effort was spent in the course of the tests to avoid the so-called cross-contamination, i.e. the samples were always handled by "clean" workers, possibly, gloves were used that were changed. The following test was performed:

- Detection of non cross-contaminated explosives
- Detection of diluted explosives with 10 % content of explosive
- Fingerprint detection
- Detection of contactless contaminated surfaces
- Detection of inert microparticles as micropreconcentrators of explosives vapours

The following samples were prepared for tests:
- *For detection of non cross-contaminated explosives*

Types of explosives:			
1.	HMX	7.	Semtex 1A + DMNB
2.	RDX	8.	PETN
3.	Permonex V 19	9.	NG SP
4.	TNT	10.	Perunit
5.	Semtex H	11.	DMNB
6.	Semtex 1A		

Sample preparation: 50 g explosive were sealed in PE sachet and inserted into 250 ml glass bottles for the tightness to be ensured.

- *Detection of diluted explosives*
Types of explosives: RDX; PETN; TNT; NG SP
The samples contained 90 % Al_2O_3 and 10 % explosives.
Sample preparation and arrangement similar to that of the first test.

- Fingerprint detection

The samples were used for contamination prepared for the first test

Types of explosives: Permonex V 19; Semtex 1A; Perunit

Arrangement of tests: 10 finger prints were applied on the surface of the suitcases using standard procedure and the ability to detect was tested

- *Detection of contactless contaminated surfaces*

Types of explosives: Semtex 1A; Perunit

Arrangement of tests: contamination of leatherette surface using standard procedure   (manipulation with the explosive above the surface)

- *Detection of inert microparticles as micropreconcentrators of explosives vapours*

50 g TNT or 50 g Perunite were placed on the bottoms of two desiccators. 30 g of inert Aerosil microparticles with a large surface were saturated in sealed desiccators with vapours of the explosive for 14 days. 3 sachets by 10 g of TNT saturated Aerosil and 3 sachets by 10 g of Perunit saturated Aerosil were prepared for tests. There were six dogs participating in tests, trained for searching of explosives (*Tab.2*).

Tab.2:

Serial No.	*Name of the dog*	Race	Origin
1.	Danny	German Shepherd	CR
2.	Adeline	Labrador	U.S.A.
3.	Bordeaux	Labrador	U.S.A.
4.	Fox	German Shepherd	CR
5.	Tom	German Shepherd	CR
6.	Art	German Shepherd	CR

Tests performance and results

- *Detection of non cross-contaminated explosives*

Eleven samples were prepared of individual mixed and plastic explosives including smokeless powder and DMNB taggant. 50 g samples were kept in ground-glass joint bottles in sealed sachets that were half an hour before tests inserted into paper boxes of dimensions 14 x 6 x 5 cm. Gradually 11 sets were prepared, consisting always of 1 box containing sample and four empty boxes in various order. Successfulness of detection by the dogs is given in *Tab.3*. Mostly two dogs only detected each set, because the tests were very demanding and the dogs could have got tired soon.

Tab.3:

Set No.	Name of the dog (Dogs No.)					
	Dany (1)	Adeline (2)	Bordeaux (3)	Fox (4)	Tom (5)	Art (6)
I. (HMX)	+	+	+	+	+	
II. (RDX)	-	(+) (+) +	(+) +			
III (Perunit)				+	+	
IV. (TNT)	+	+				
V. (Semtex H)			+	+		
VI. (Semtex 1A)	+				+	
VII. (Semtex 1A + DMNB)		+	+			
VIII. (PETN)				+	+	
IX. (NG SP)	+	-				
X. (Perunit)			+	- +		
XI. (DMNB)			+		-	

Note: +   detection
      -    miss
    (+)   mistaken detection

- *Detection of diluted explosives with 10 % content of explosive*

Four samples were prepared of diluted explosives by blending of Al_2O_3 (90 % by weight) and explosive (10 % by weight). The samples in sealed sachets were kept in ground-glass joint bottles and a half an hour before the tests inserted into paper boxes similarly as in the previous test. Four sets were formed, each set consisted of 5 boxes, of which 4 boxes being empty and 1 containing the sample. Results of explosives detection are given in *Tab.4*.

Tab.4:

Set No.	Serial No. of the dog					
	1	2	3	4	5	6
I. (RDX)			(-) (-) +	+		
II. (TNT)		+	-			+
III. (PETN)	+				(-) +	
IV. (NG SP)		(-) +			+	+

- *Fingerprint detection*

The samples of explosives from the first tests were used for these tests, namely Permonex V19, Semtex 1A and Perunit. Into these explosives thumb was always pressed several times using relatively great force and by means of that thumb 10 fingerprints were made on the suitcase lid and on the area around suitcase handle. The suitcase was placed into garage among other objects and dogs were to seek out contaminated suitcase. Results are given in *Tab.5*.

Tab.5:

No. of suitcase	Explosive	Serial No. of the dog					
		1	2	3	4	5	6
1	Permonex				the dog perhaps tired, confused		+
2	Semtex 1A			- +		+	
3	Perunit		+		+		

- *Detection of contactless contaminated surfaces*

The samples from the first tests were used for these tests again, namely Semtex 1A and Perunit. Some 20 cm above the leatherette of dimensions of ca 30 x 30 cm about 50 g explosive were rolled and partitioned carefully to avoid greater particles to fall down directly on the leatherette. The leatherette was then placed in the room with sitting suite to be sought out by dogs. Results are given in *Tab.6*.

Tab.6:

Explosive	Serial No. of the dog					
	1	2	3	4	5	6
Semtex 1A		+			+	
Perunit				+		+

- *Detection of inert microparticles saturated with explosives vapours*

14 days prior to planned tests term, 50g TNT or 50 g Perunit were placed on the bottoms of two desiccators. Above those explosives, Aerosil (very fine SiO_2) was placed on plates. After 14 days that Aerosil was weighed out by 10 g, sealed into PE sachets and inserted into ground-glass joint bottles. At tests, TNT sample was placed on the loading area of a delivery van to be sought out by the dog, Perunit sample was placed in lavatory and into half closed drawer in an office. Results are given in *Tab.7*.

Tab.7:

Explosive	Placement	Serial No. of the dog					
		1	2	3	4	5	6
TNT	Loading area of delivery van			-			+
Perunit	Lavatory		+		+		
Perunit	Drawer in an office		+	+			

After tests, saturation (contamination) of the Aerosil samples was verified by means of detection set DETEX. At TNT, slight black brown and green (dinitro?) dots appeared on the witnessing filter test paper, at Perunit, the presence of nitroester was proved by a very distinct pink color. The content of TNT and 2,4 DNT respectively NG and EDGN was determined by means of quantitative analysis in ppm with the following results:

Explosive	TNT	Perunit
TNT	4,5	
2,4 DNT	4,5	
NG		22
EGDN		1200

CONCLUSIONS

General conclusions from the VÚPCH workers viewpoint are as follows:

- In general - unexpectedly good results of detectability of explosives by means of dogs (when compared with detectors efficiency tests performed in VÚPCH), especially the dog No. 6 was 100 % successful, but the other dogs' performance, however, was on a very good level too, if we consider the amount and time consuming nature of the work performed.
- In our opinion, seeking out hidden charges based on explosives vapours is more effective for detection of explosives by means of dogs than that based on particles, which is, in fact logical with regard to olfactory organ of a dog.
- It follows from the results of the tests of non cross-contaminated explosives that dogs are able to seek out all commonly used explosives, only a small hesitation at seeking out nitroglycerin smokeless powder and Perunit was rather surprising, considering a high vapour tension of these nitroesters. On the contrary, seeking out plastic explosives taggant as such (DMNB is not an explosive) by means of the dog trained in the U.S.A. perhaps witnesses of training for this compound.
- Also relatively successful seeking out the samples containing only 10 % of active substance is very satisfactory because it will make it possible in the future to supply such samples for dogs' training that will not be necessary to be stored in the regime of explosives and, at the same time, the samples will be of parameters required for training (we count with the content of about 30% active substance in the sample).
- Fingerprint detection was rather more difficult for the dogs, but it was relatively well managed as well. There are, however, indications there, that for these purposes the use of dogs is not probably the optimum solution.
- The dogs managed the detection of contactless contaminated leatherette surface without any hesitation, there is likely to be present higher amount of explosive.
- At detection of contaminated inert microparticles, very high sensitivity of dog's olfactory organ was proved, because even the sample was found with TNT vapours. After the test with DETEX kit it is possible to reckon, that TNT content in the sample is of about 10^{-6} to 10^{-7}g, with Perunit maybe a little higher. If we take into consideration that the sample is sealed in PE sachet, the efficiency of dog's performance to detect of explosive vapour is admirable indeed.
- In conclusion it is possible to state that the results of the tests brought along some important information (e.g. for preparation of samples for training) and that, on the basis of the results achieved, it would be useful to go on in co operation.

REFERENCES

[1] Yinon Jehuda: Forensic and Enviromental Detection of Explosives, Weizmann Institute of Science, Rehovot, Israel, 1999

[2] Malotky L.O : Research and operational deployment of explosive detection dogs, ICAO, Ad Hoc Group of Specialists, Twelfth Meeting, Montreal, 2000

[3] Strobel R.A. et al.: The ATF Canine explosives detection program, ICAO, Ad Hoc Group of Specialists, Eighth Meeting, Montreal, 1994

[4] Strobel R.A., Noll R.: Nitromethane K-9 Detection Limit, 7th International Symposium on the Analysis and Detection of Explosives, Edinburgh, 2001

[5] Biederman G.B.: Vapour preconcentration in the detection of explosives by animals in an automated setting, Advances in Analysis and Detection of Explosives, Kluwer Academic Publishers, Dordrecht, 1993

[6] Groznov I.N.: On the experience of statistical estimation of quality of a large group of dogs training for explosives detection, ICAO, Ad Hoc Group of Specialists, Seventh Meeting, Montreal, 1993

[7] Bouisset J.F.: Statistical Measurement of the Effectiveness of the Detection of Vapours and Traces of Explosives using Dog handling Teams – ICAO, Ad Hoc Group of Specialists, Seventh Meeting, Montreal, 1993

[8] Guicheney G.: Comparison of the performance of an automatic trace detector and that of dog handling teams – ICAO, Ad Hoc Group of Specialists, Eleventh Meeting, Montreal, 1997

[9] Göth Ann et al: How do dogs detect landmines – GICHD, Geneva, Marc

NEW DRIFT SPECTROMETER WITH THE SURFASE IONIZATION OF THE ORGANIC MOLECULES

V.I. KAPUSTIN

Join-stock Company «GYCOM» Obrucheva str., 52 Moscow, 117393, Russia. E-mail: gycom_m@orc.ru

INTRODUCTION

Ionization of organic molecules on the heated solid under vacuum is a well known physical phenomenon [1 – 2]. The experimental data were usually interpreted in terms of the known Saha-Langmuir equation.

In the mid-1960s, selective surface ionization of some types of organic compounds (amines) on a heated pre-oxidized refractory metals in the ambient air was discovered [3 – 4]. It was assumed that the surface ionization occurs in two steps – dissociation of organic molecules on the adsorbent surface and the following desorbtion of the fragments in accordance with Saha-Langmuir equation [5 – 6]:

$$I_i (T) = evS\gamma_i (T) \beta_i (T, E) \tag{1},$$

where T is temperature, E is the electric field strength near the solid surface, $I_i (T)$ ia the current of ions of the i --the type from the surface, e is the electron change, v is the stream of ions per unit surface of the solid, S is the surface area of the solid, $\gamma_i (T)$ is the correction factor, which has been called [5] the coefficient of conversion of the stream of organic molecules into species of the i – the type, $\beta_i (T,E)$ is the surface ionization factor for type I - the species, equal to [5]:

$$\beta_i(T) = \cfrac{1}{1+\cfrac{1}{A_i(T)} \exp\left(\cfrac{V_i - \varphi - (eE)^{1/2}}{kT}\right)} \tag{2},$$

where V_i is the adiabatic ionization potential for i – the particles, $A_i (T)$ is the ratio of the statistic sums for the neutral states of i – the species, φ is so-called "ionic work function" for the solid.

The traditional approach to the surface ionization of organic molecules at air conditions does not describe adequately this physical phenomenon, because it requires introduction of the notion of "ionic

M. Krausa and A. A. Reznev (eds.),
Vapour and Trace Detection of Explosives for Anti-Terrorism Purposes, 79-86.
© *2004 Kluwer Academic Publishers. Printed in the Netherlands.*

work function", which has not definite physical meaning; cannot interpret the experimentally observed of the ion current versus the electric field strength; does not account for the presence of $(M + H)^+$, $(M - H -2nH)^+$ and $(M - R -2nH)^+$ ions in the spectra; cannot account for the "concentration dependence" of the ion current, i.e., the dependence of the magnitude of ion current on the flux of organic molecules onto the solid surface; and does not actually explain the selectivity of ionization of organic amino molecules on oxidized surfaces of refractory metals.

PHYSICAL MODEL OF THE SURFACE IONIZATION

When considering the physical model of the surface ionization of amines and nitro compounds organic molecules, one should take into account the electronic structure of the amino and nitro groups contained in the given compounds and the structural features of the metal oxide surface.

Fig. 1 shows the model of the oxide surface containing so-called Brönsted acid sites {BAS}1 and Brönsted base sites {BBS}2 [7]. They are formed on the surfaces of W, Mo, Re, Al, Zr, Mg , and other oxides and are represented by hydrogen ions and hydroxyl groups chemisorbed on the oxygen and metal atoms of the oxide, respectively, following the dissocialize adsorption of water molecules on the oxide surface. The same figure shows a surface alkali metal ion {AMI} 3 for the oxide having a complex composition, e.g., for the surface of an alkali metal oxide bronze [8]. The oxide surface containing active sites of the {BAS}, {BBS} and {AMI} types can initiate surface reactions with the exchange by protons, alkali metal ions and hydroxyl ions.

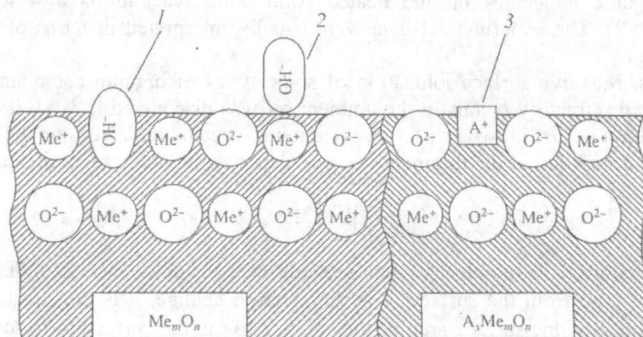

Fig. 1: Model of the surface of Me_mO_n oxide (on the left) and $A_xMe_mO_n$ alkaline oxide bronze (on the right): 1 – Bronzed said site {BAS}, 2 – Bronzed base site {BBS}, 3 – surface alkali metal ion {AMI}. Me means a transition metal.

The nitrogen atom of an organic amino carries a lone pair of valence electrons [9], which can add a proton; this gives a secondary ion and a closed electron shell at this proton:

$$R_1 : N : \quad + H^{\oplus} \rightarrow \quad [\ R_1 : N : H\ ]^{\oplus} \tag{3}$$

with R_2 (lone pair) above and R_3 below on both sides.

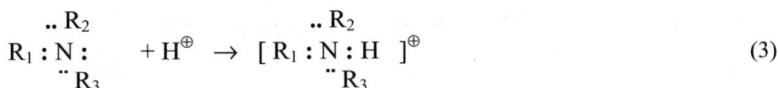

In the molecule of an organic nitro compound, the nitrogen atom is bound to an oxygen by a so-called semi polar bound, which has a substantial dipole moment [9]; an alkali metal ion or a proton can add by the reaction:

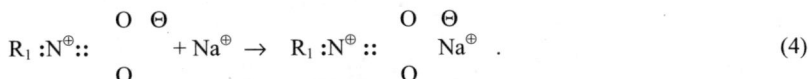

$$R_1 : N^{\oplus} :: \quad + Na^{\oplus} \rightarrow \quad R_1 : N^{\oplus} :: \quad Na^{\oplus}\ . \tag{4}$$

with O / $\ominus$ / O groups on both sides.

The stability of the secondary ion being determined by the ionic radius of the corresponding alkali metal or the proton.

Thus, amines and nitro compounds can form ions on the oxide surface without electron exchange between the organic molecules and the surface. The whole process can be represented by a sequence of ionization reactions:

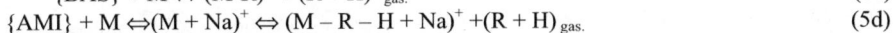

$$\{BAS\} + M \Leftrightarrow (M+H)^+ \Leftrightarrow (M-H)^+ + H_{2\ gas.} \tag{5a}$$
$$\{BBS\} + M \Leftrightarrow (M + OH)^- \Leftrightarrow (M-H)^- + H_2O_{\ gas.} \tag{5b}$$
$$\{BAS\} + M \Leftrightarrow (M-R)^+ + (R + H)^+_{\ gas.} \tag{5c}$$
$$\{AMI\} + M \Leftrightarrow (M + Na)^+ \Leftrightarrow (M - R - H + Na)^+ + (R + H)_{\ gas.} \tag{5d}$$

Since the Saha-Langmuir equation is implicated to reactions (5a) - (5d), the rate of surface ionization of organic molecules (ion current) can be calculated for these reactions using, for example, methods of the theory of absolute reactions rate:

$$I_i(T) = A \frac{P^n}{T^{5/2}} \exp(-\frac{\Delta E}{kT}) \tag{6},$$

where A is a constant, P is the partial pressure of the organic molecules near the oxide surface, ΔE is the activation energy for ion desorption from the oxide surface, n is the reaction order determined by the order of the bonds between the adsorbed organic molecules and the oxide surface.

SURFACE IONIZATION OF AMINES

Fig. 2 shows the temperature dependences of the background current emitted by an oxidized molybdenum surface in the ambient air for positive (curve 1) and negative (curve 2) ions. Peak 1 of the ion current corresponds to the desorption of hydroxyl ions from the {BBS} sites, peak 2 shows the proton desorption from the {BBS} sites, peak 3 shows the proton desorption from the {BAS} sites, peaks 4 are related to a phase transition in the molybdenum oxide layer, and peaks 5 correspond to the thermal vibrations of ions on the oxide surface. The calculation of the desorption activation energy for peaks 1, 2 and 3 resulted in values of 1.27, 1.36, and 2.34 eV respectively.

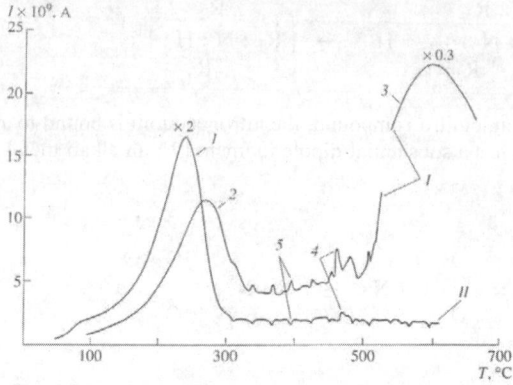

Fig.2: Thermogramms of the background current of ions emitted by the oxidized Mo surface in ambient air.

Fig.3 shows the experimental dependence of the positive ion current induced by the surface ionization of Novocain molecules on the Novocain dose introduced in the ionization zone.

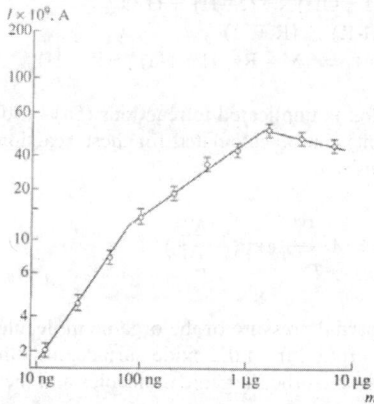

Fig.3: Positive ion current vs. the dose of organic molecules after ionization of Novocain on the oxidized molybdenum surface in the ambient

The concentration dependence has three characteristic regions differing in the slope angle and, hence, in the reaction order of ion desorption. This value and the activation energy of desorption, which we found from the temperature dependences of positive ion current for these regions, are shown in the *Tab. 1.*

Tab.1: Parameters of ionization of Novocain molecules on the oxidized molybdenum

Novocain dose in an air stream, ng	Activation energy of ion desorption, eV	Ionization reaction order, rel. u.	Ionization efficiency, C/mol
3300	1,77	- 1/6	20
400	1,29	1/2	100
50	1,26	1	200

SURFACE IONIZATION OF NITRO COMPOUNDS

Fig.4 shows the temperature dependence of vaporization of sodium atoms from the sodium oxide bronze under high vacuum (curve 1) and the temperature dependence of the background current of the sodium positive ions emitted by the same oxide bronze in the ambient air (curve 2). The same Figure shows the temperature variation of the positive ion current resulting from ionization of organic nitro compounds: trinitrobenzene (curve 3), octogen (curve 4), tetraethylnitrile (curve 5), and trinitrotoluene (curve 6) on the oxide bronze surface in the ambient air. The maxima of the molecular ionization efficiency are well correlated with the maxima of sodium ion concentration on the surface of the oxide bronze, which is consistent with the model views outlined above.

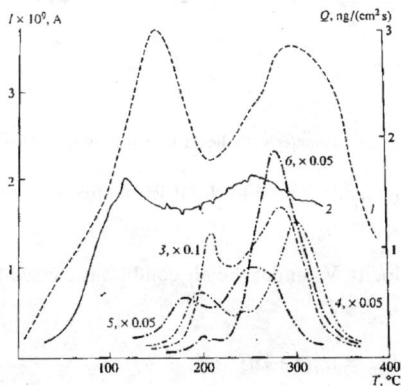

Fig.4: Temperature dependences of (Q) vaporization of Na atoms from the sodium oxide bronze, (1) background current of the sodium positive ions, and positive ion current upon ionization of (3) trinitribenzen, (4) octogen, (5) tetraethylnitrile, and (6) trinitrotoluene molecules on the surface of the sodium oxide bronze.

DRIFT SPECTROMETER WITH THERMAL SURFACE IONIZATION OF THE ORGANIC MOLECULES

Thermal surface ionization drift spectrometer is the new variety of devices known as ion non-linear drift spectrometer, field ion spectrometer, ion mobility increment spectrometer and usually used for drug and explosive detection [11-12]. The organic molecules ionization in all known devices is realized by using of radioactive Ni^{63} or H^3 [13-14].

Fig. 5 shows the design of drift spectrometer [15]. The main part of device is the flat ion emitter 1 with bib surface area. Organic molecules enter into device with the air flow Q_1 and then selectively ionized on the heated surface of the emitter. Ion emitter (for amines) is made on the base of the oxide transition metal or (for nitro compounds) on the base of the sodium oxide bronze. The heater 2 controlled the temperature of the emitter (200 - 500 ^0C). The main transport gas flow Q_0 is also the ambient air.

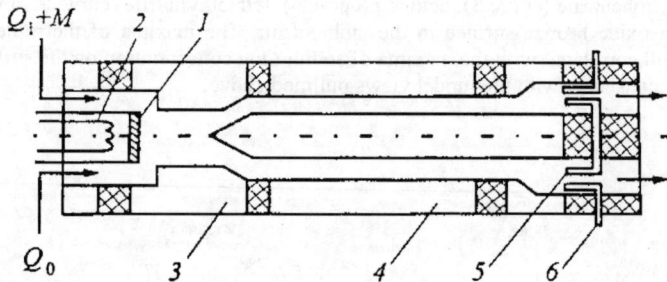

Fig.5: Design of drift spectrometer with the surface ionization of organic molecules.

Ion lens 3 forms the ion flow and sends ions into the drift analyzer 4. Around the ions collector 6 is situated the suppressor 5.

It is known [16] that ion drift velocity V_d under the air conditions caused by an action of electric field E:

$$V_{\mathcal{A}} = \mu_0 (1 + \alpha E^2)E \qquad (7),$$

where μ_0 is the drift mobility, α is the nonlinear part of drift mobility. Due to our new type of the ion drift spectrometer it is possible to measure the presence of amines or nitro compounds in gas flow Q_1 (emitter current) and type and quantity of this organic molecules (collector current). As an example, on *Fig. 6* is presented the drift-spectra of Hinin (next test amines). Its concentration in gas flow Q_1 was equal $1*10^{-10}$ gramm/sm^3. Drift-spectra of all investigated amines contain two peaks with two different values of α, presented in *Tab. 2*.

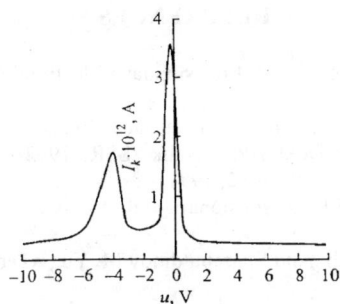

Fig.6: Drift-spectra of Hinin: its air concentration is equal $1*10^{-10}$ g/sm^3.

Tab.2: Non-linear part of ion drift mobility.

Peak in	α, $*10^{10}$, sm^2/B^2				
Spectra.	Novocain	Papaverin	Dimedrol	Bencain	Hinin
Main	+ 0,018	- 0,045	-.0,034	+ 0,034	+ 0,023
Next	- 0,65	- 0,21	- 0,53	- 0,50	- 0,94

The sensibility of device is equal $1*10^{-16}$ gram (probe) or $3*10^{-17}$ gramm/sm^3 (concentration) by emitter current and $1*10^{-12}$ gram (probe) or $3*10^{-13}$ gram/sm^3 (concentration) by collector current.

CONCLUSIONS

1. We observed experimentally for the first time a new physical phenomenon, namely selective ionization of organic molecules (nitro compounds) on the surface of some type of alkali metal oxide bronze in the ambient air. A physicochemical model for this phenomenon was proposed, which take into account the molecular structure of nitro compounds and the surface structure of oxide bronzes.
2. It was shown experimentally and substantiated theoretically for the first time that the selective ionization of organic molecules of amines on the oxidized surface of transition metals in the ambient air is due the interaction of amino groups of these molecules with the Bronsted acid and base sites, which exist on the oxide surface as hydrogen and hydroxyl ions from dissocialize adsorption of water molecules on this surface.
3. New type of drift spectrometer with the surface ionization of the organic molecules under air conditions was developed.
4. The sensibility of device is equal $1*10^{-16}$ gram (probe) or $3*10^{-17}$ gram/sm^3 (indicated mode by measuring emitter current) and $1*10^{-12}$ gram (probe) or $3*10^{-13}$ gram/sm^3 (identification by measuring collector current).

REFERENCES

1. Dobretsov L.N. and Gomoyunova M.V. Emissionnaya Electronika, 1966, "Nauka", Moscow, Russia.
2. Rasulev U. Kh., Zandberg E. Ya., «Progress in surface science» . 1988, Vol. 28, 3/4, p. 181 - 212.
3. Zandberg E.Ya. and Ionov N.I., Dokl. Akad. Nauk SSSR, 1962, v. 141, pp. 139- 142.
4. US Patent № 5028544, G 01N33/00, Jul. 2, 1991.
5. Nazarov E.G. and Rasulev U. Kh., Nestatsionarnye protsessy poverkhnostnoi ionizatsii, 1991, "Fan", Tashkent, Uzbecistan.
6. Zandberg E. Ya, Rasulev U. Kh. And Sharaputdinov M. Yu., «Teor. Eksp. Khim.» (Rus.), 1971, v. 7, pp. 363 – 389.
7. Morrison S.R., The chemical physics of surface. N-Y, Plenum Press, 1977.
8. Oksidnye bronzy, Spitsyn V.I., Ed., 1982, "Nauka ", Moscow, Russia.
9. Potapov V.M., Organicheskaya khimiya, 1970, "Prosvyshenie", Moscow, Russia.
10. Lushpa A.I., Osnovy lhimicheskoi termodinamiki i kinetiki khimicheskikh reaktsii, 1981, "Mashinistroenie", Moscow, Russia.
11. Kolla Peter , «Detecting hidden explosives», Analytical Chem., 1995, v.67, no.5, March 1.
12. «Explosives Detecting Systems», Review of the In Vision Technologies Inc. (US), ICAO Journal, 1995, Dec., p. 11-13.
13. McGann W., A New, High Efficiency Ion Trap Mobility Detection System For Narcotics, Proceedings of SPIE, 1996, v. 2937, p. 78 - 88.
14. Carnahan B., Day S., Kouznetsov V., Tarassov A., Field Ion Spectrometry - A new technology for cocaine and heroin detection. Proceedings of SPIE, 1996, v. 2937, p. 106 - 119.
15. Kapustin V.I. and Boron A.A., Physics of electronic materials, Int. Conf. Proc., Kluge, Russia, October 1 – 4, 2002, p.p. 365 and 393.
16. McDaniel E., Mason E., The mobility and diffusion of ions in gases. John Wiley and Sons, NY, 1973.

HIGH-SPEED GAS ANALYSIS FOR EXPLOSIVES DETECTION

V.M. GRUZNOV, M.N. BALDIN, V.G. FILONENKO

The Design & Technological Institute of Instrument Engineering for Geophysics and Ecology (IDE), the Siberian Branch of RAS e-mail: majak@uiggm.nsc.ru

INTRODUCTION

Detection of explosives and drugs is an urgent problem in the world today for the purposes of anti-terrorism checking, clearing the areas of military conflicts of mines, investigating the cases of explosives and drugs usage, including identification of people after their contact with explosive devices and drugs. The most attractive among physical methods are the techniques based on the detection of explosives and drugs by "odor" like a trained dog.

A physical foundation for explosives detection by "odor" is explosive vapor detection. Methods of gas analysis offer advantage over the well-known methods of explosives detection (radiowave, induction, nuclear-and-physical, nuclear quadrupole resonance). This advantage lies in the fact that during the detection the tested objects are not exposed to penetrating radiation, which in some cases can trigger an explosive device.

The report briefly outlines the technologies of high-speed gas chromatography, ion mobility increment spectrometry, chromato-mass-spectrometry for explosives detection developed by the Design & Technological Institute of Instrument Engineering for Geophysics and Ecology (IDE), the Siberian Branch of RAS.

GAS CHROMATOGRAPHY

The work experience has shown that for fast detection of explosive vapor traces one can use gas chromatographs (GC) with the following parameters: separation efficiency, 1000÷3000 theoretical plates; separation time, 10-180 s; detectors: electron capture, molecular condensation nuclei detector, ion detector with a varying selectivity. The processes that take place in the gas chromatographs are given below.

M. Krausa and A. A. Reznev (eds.),
Vapour and Trace Detection of Explosives for Anti-Terrorism Purposes, 87-99.
© 2004 *Kluwer Academic Publishers. Printed in the Netherlands.*

High-speed Separation

High-speed separation in 10-180 s is achieved through multicaplillary columns (MCC). A multicapillary column is a 1 m long monolithic bundle with about one thousand capillaries 40,0 μm in ID. At present MCC coated with different liquid phases, like SE-30, SE-54, Carbowax -20M, СТКФТ-50X, OV-61 and etc. are available. MCC allow to: a) analyze samples with a high content of the compound detected (500 $\div$3000 ng); b) to separate compounds at a lower temperature, for example, explosives can be separated at 150$\div$170° C, instead of 220° C. Low separation temperature decreases power consumption of a self-contained device. GC designed for explosives detection use short multicapillary columns with a length of 22 cm and a separation efficiency of 2000$\div$3000 theoretical plates. A carrier-gas flow rate of 30$\div$60 cm^3/min in these MCC is achieved through a low pressure fall in the range from 0.35 to 0.7 atm. The time of explosives separation does not exceed 1 min. MCC are given on *Fig.1*. The main parameters of the MCC can be expressed by:

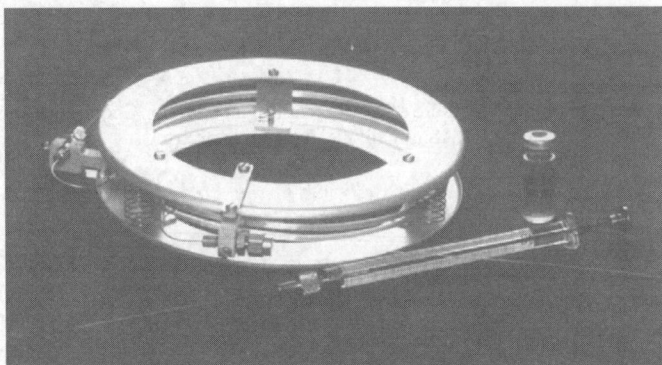

Fig. 1(a): General view of a multicapillary column (mcc)

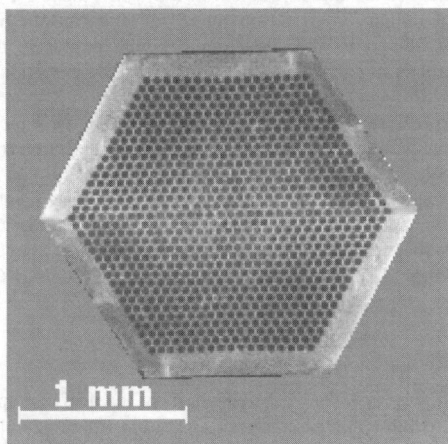

Fig. 1(b): MCC in cross-section (MCC is a monolithic bundle with about one thousand 40 μm ID capillaries with immobilized liquid phases that separates compounds in parallel)

$$N = \frac{L}{H_0 + (\frac{\delta_s}{S_0})^2 L} \tag{1}$$

where:

δ_s - is a root-mean-square deviation of cross-sectional area of capillaries
S_0 - is an average capillaries area in a bundle
N – is a number of theoretical plate

$$N_{max} = \frac{1}{(\frac{\delta_s}{S_0})^2} \tag{2}$$

where:

L – is a column length,
H_0 – is a height of equivalent theoretical plate for the capillary column with a capillary area equal to S_0.
Fig. 2 shows a high rate of explosives separation by the MCC, where TNT isomers are separated in 20 s.

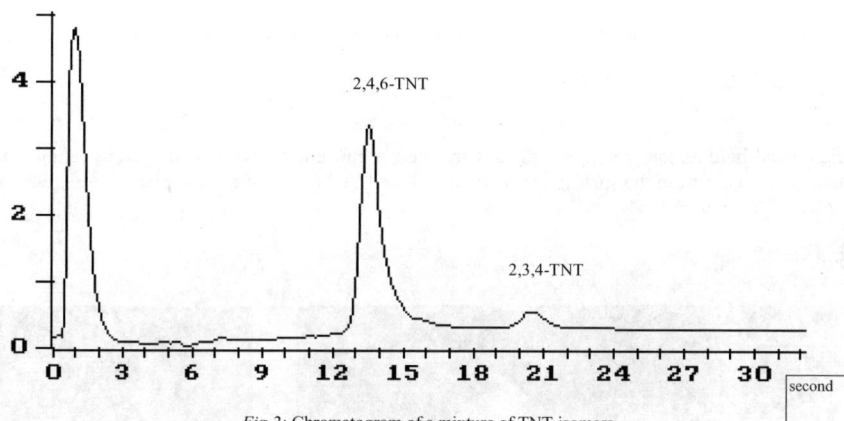

Fig.2: Chromatogram of a mixture of TNT isomers

Remote Explosive Vapor Sampling

For remote explosive vapor sampling from an object surface a vortex sampling technique is developed. The technique consists in blowing an object with a whirled air stream which forms an air flow direct into a concentrator. The flow contains explosive vapors. For efficient sample transport a relationship between the whirled air flow and the air flow drawn off the whirl site is optimized. The vortex sampler is represented schematically on *Fig.3*.

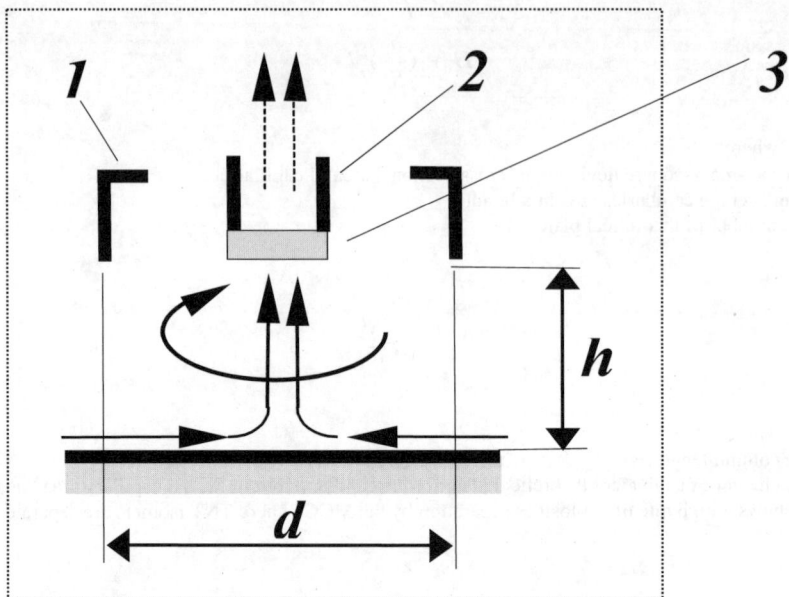

Fig.3: Diagram of a vortex sampler of EKHO-M gas chromatograph: 1 –vortex chamber; 2 –air suction line; 3-concentrator

Efficient hand-held remote vortex samplers with a power consumption of 5 W that allow remote sampling at a distance of 0-20 cm from the surface tested are developed. Examples of the sampler use are given on *Fig. 4.*

Fig.4: Examples of air sampling on detecting explosives

High-speed VaporTtrapping

A high-speed technology of explosive vapor trapping on a concentrator is developed. The concentrator is made of n thin channels with diameter d and length ℓ. The concentrator efficiency is characterized by breakthrough, β, that is determined as a ratio between the amount of a compound passed through the concentrator and the amount that entered the input, and is calculated from the formula (3):

$$\beta = exp\ (-Q_d/Q) \tag{3},$$

where Q is an air flow rate through the concentrator, $Q_d = 6\ \pi Dn\ell s$, D is a coefficient of explosive vapor diffusion, s is a probability of explosive molecules adhesion to the concentrator surface.
A mode of high-speed trapping is given by air flow rate, Q, such that a high increment of the detector response signal is obtained. To choose Q, the dimensionless Q,- dependence of the response signal is given:

$$A' = Q/Q_d\ [1 - exp\ (-Q_d/Q] \tag{4}$$

From (2) follows that Q should be chosen almost equal to Q_d.

Based on the analysis of the high-speed thermal desorption, efficiency, $\mathcal{E}_i$, of the concentrator sample injection versus sample injection time, t_i , and time of the concentrator heating, t_d, is determined.

$$\mathcal{E}_i = [1 - exp(-t_d/\alpha)] \cdot [1 - exp(-t_i/b)] \tag{5},$$

where α and b are constant values.
Fig. 5 shows experimental and theoretical curves of the efficiency of sample injection versus injection time calculated from (3). One can see that with an injection time of 0,5 s the efficiency is more than 0.5.

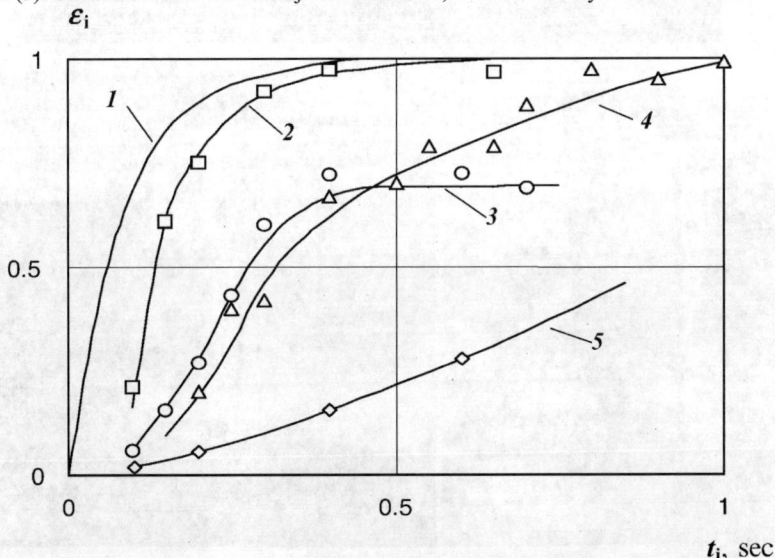

Fig.5: Efficiency of sample injection versus injection time: 1=complete instantaneous desorption (estimated curve); 2=injection of mononitrotoluene with the heating of a concentrator-trap from non-modified stainless steel grid; 3=2,4,6-TNT injection with the heating of a concentrator- trap made from non-modified multichannel glass plate 1 mm thick with channels 40 μm in diameter; 4= 2,4,6 -TNT injection with the heating of a concentrator- trap from non-modified stainless steel; 5= 2,4,6-TNT injection without heating a concentrator- trap from non-modified stainless steel grid.

EKHO Gas Chromatographs

Modes of high-speed sampling, sample injection and separation are realized in portable high-speed gas chromatographs of the EKHO-series designed for explosives detection and ecological monitoring. An EKHO-M device is being used by the Ministry of Interior, Russia as an explosive detector. All GC of the EKHO series are provided with a built-in microprocessor for setting GC parameters, processing the results of analysis. The device is connected to IBM PC and is provided with an automatic sampler for analyzing gaseous samples with an interval of 1.5 min and more. The EKHO-M device is shown on *Fig. 6.*

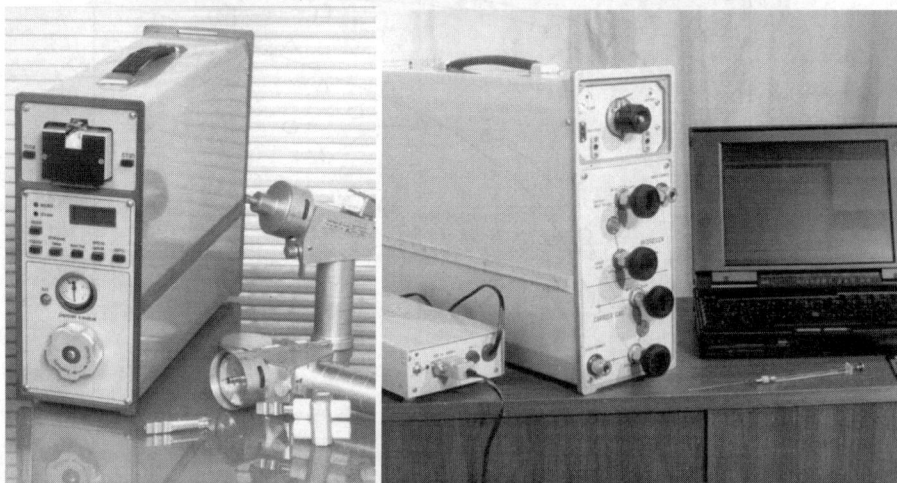

Fig.6: EKHO gas chromatographs:
1-EKHO-M with a syringe injector suited for the USA market ; 2-EKHO-M with a concentrator injector; 3- EKHO- FID with a syringe injector; 4-vortex sampler; 5-concentrator injector; 6-syringe injector; 7- power source; 8 -PC

Specifications of the EKHO-M are:

detection limit for 2,4,6 TNT - $10^{-14}\,g/\,cm^3$

time of 2,4,6 TNT vapor analysis -   15 s

time of a concentrator sampling -   15....60 s

The latest EKHO-M modification is an EKHO-A gas chromatograph with air as a carrier gas, which makes the device more convenient in operation under field conditions. The possibilities for the MCC operation with air as a carrier-gas are shown. An ion detector with a varying selectivity and a filter for air purification are developed.

The EKHO-A is shown on *Fig.7*. The main device characteristics are given below.

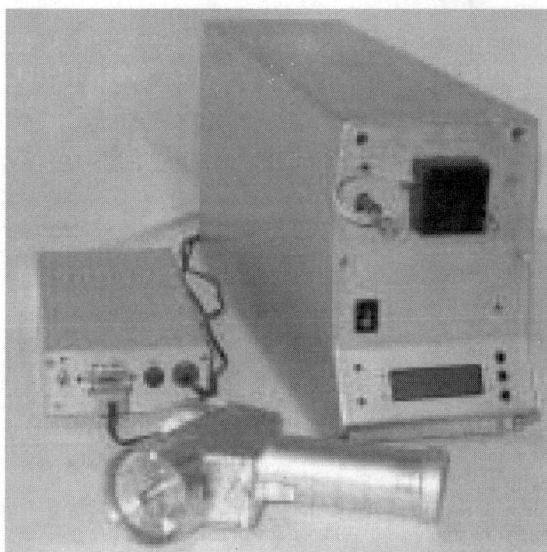

Fig.7: EKHO-Air Portable High Sensitive Gas Detector-Analyzer with Air as a Carrier Gas

SPECIFICATION:

Detector - ion detector with a varying selectivity;

Gas chromatographic column - high-speed multicapillary column 20 cm in length (efficiency, theoretical plates/m - 8000-10000, temperature, °C - 40÷ 190, precision of temperature maintenance, °C - not worse than 0,2; coating - OV-624, SE-30, SE-54, Carbowax-20M);

Sample injector:

a) syringe injection,
b) concentrator injection with a sorbent volume of $0.01 \div 0.18$ cm^3,
c) automatic sample loop.
Operating temperature range, ^0C - +5 to 40, up to +50
Power consumption, W - no more than 70
Dimensions, mm - length - 500;width: 135;height: 330
Weight without accumulator, kg - 6
Weight with accumulator, kg - 9
Sampler weight, kg – 1

Detection limit of a chromatograph:

Detection limit, mg/m^3 analysis time

Direct air sample injection

	Detection limit, mg/m³	analysis time
Sarin(GB)	0,015	60 s
Soman(GD)	0,03	60 s
V- gases (VX)	0,15	70 s
mustard (HD)	0,2	70 s
Lewisite(L)	0,6	30 s

Syringe injection with a syringe volume of 1 mcl

Sarin(GB)	$6,0*10^{-6}$	60 s
Soman(GD)	$1,2*10^{-5}$	60 s
V- gases (VX)	$4,2*10^{-5}$	70s
Mustard(HD)	$7*10^{-5}$	70 s
Lewisite(L)	$2,3*10^{-5}$	30 s

With concentrator-trap, mg/m^3:

Sarin(GB) in air sample with a volume of 2 l - not worse than $1,5*10^{-5}$
Soman (GD) in air sample with a volume of 4 l - not worse than $2*10^{-5}$
V-gases (VX) in air sample with a volume of 20 l - not worse than $9,5*10^{-6}$
Mustard(HD) in air sample with a volume of 4 l - not worse than $8,4*10^{-5}$
Lewisite (L) in air sample with a volume of 1 l - not more than $2,5*10^{-4}$
Trinitritoluene (TNT) in air sample with a volume of 1 l - not worse than $9*10^{-15}$
Computer for setting GC parameters, processing the results of analysis.
Compound identification by the available database and operator warning on the presence of the searched compounds by means of the audible and visible alarms.

Power supply:

- embedded accumulator 12V, 12 Å*hour (time of device operation from the accumulator at GCC temperature of 100ïÑ, from turning on not less than 3,5 h);
- external DC source (output voltage from 11 to15 V; rated power not less than 70 W; can operate from a car cigarette lighter or accumulator of 12 V);
- external AC source - 220/115 V, 50- 60 Hz.
 Fig.8, 9 present chromatograms of different explosives, separation time is under 30 s.

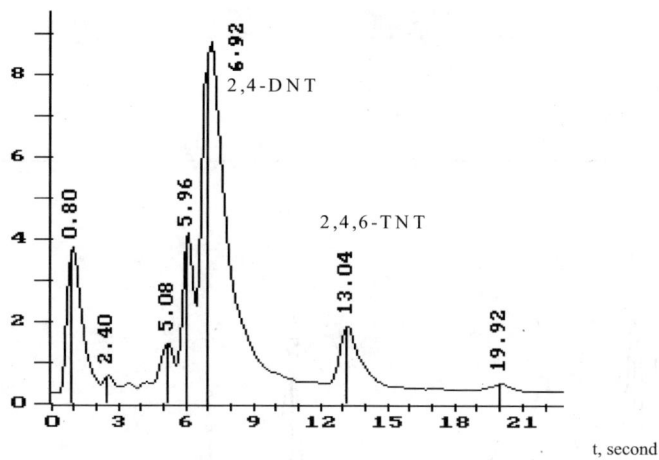

Fig. 8(a): Chromatogram of commercial TNT

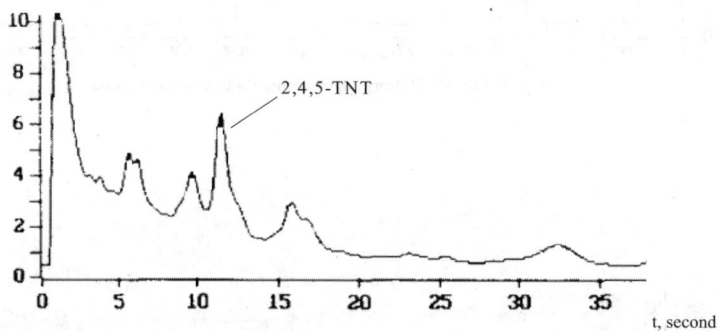

Fig. 8(b): Chromatogram of an air sample from a surface of a pocket containing an explosive cartridge

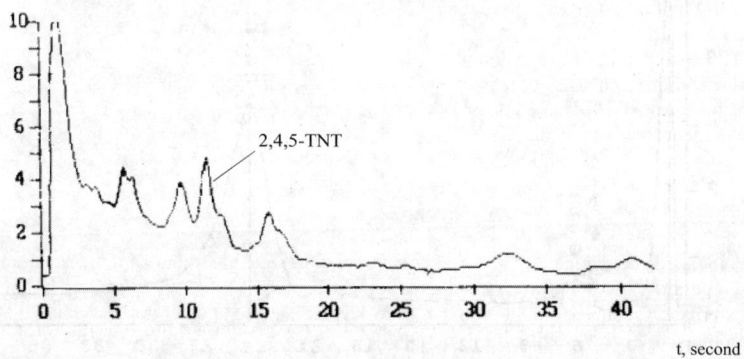

Fig. 8(c): Chromatogram of an air sample from a surface of a pocket from which an explosive cartridge was taken out

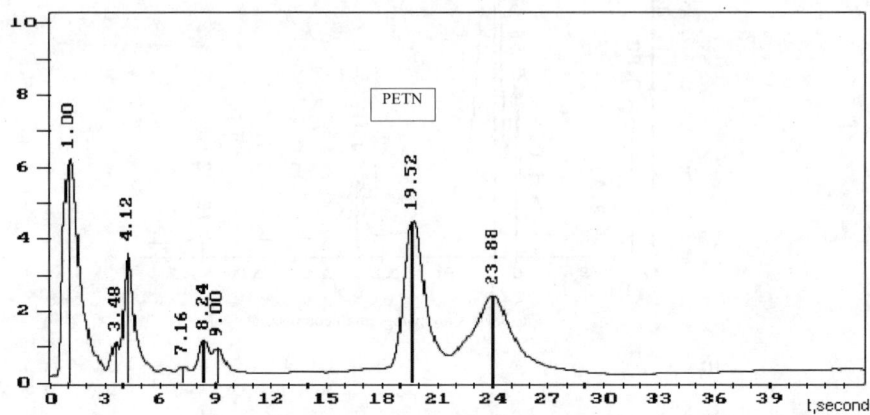

Fig. 9(a): Chromatogram of PETN from a Bickford fuse

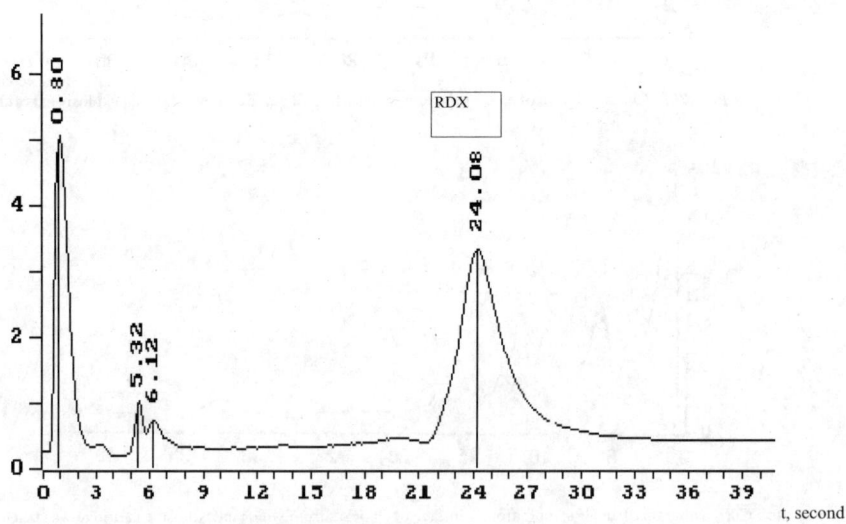

Fig. 9(b): Chromatogram of RDX

To reduce the detection limit for explosives to 10^{-15}g/cm^3, a concentrator with an enlarged diameter (to 40 mm) provided with a system of a sample reconcentration in a cooled capillary (reconcentration sample injector) is used. Chromatograms obtained with the reconcentration sample injector (RSI) are given on *Fig. 10*.

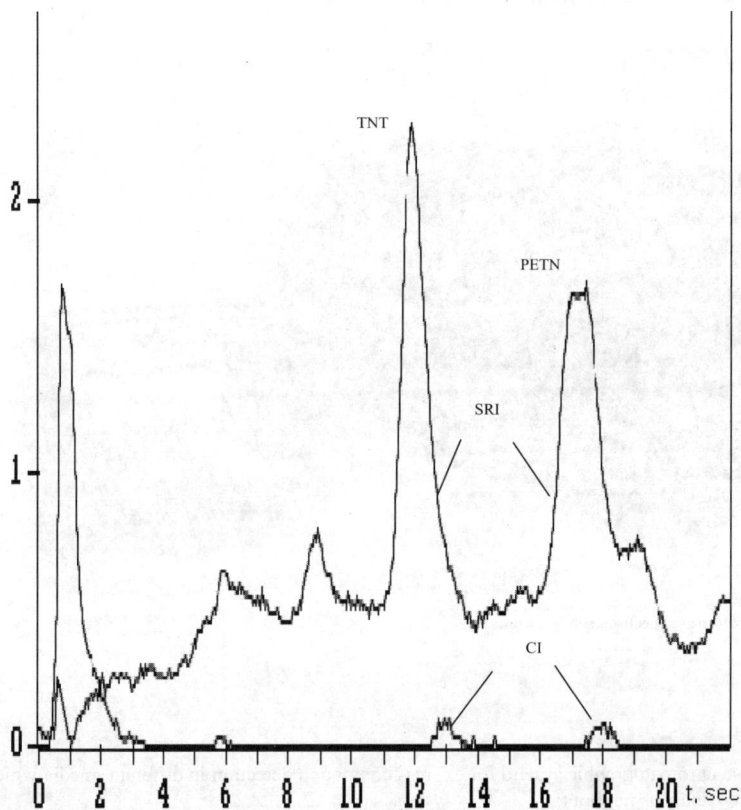

Fig.10: Chromatograms obtained with the EKHO-M/CI and EKHO-M/SRI. Air samples were taken with a vortex sampler and above the source of a mixture of TNT and PETN vapors at a distance of 100 mm.

2. MOBILE CHROMATO-MASS-SPECTROMETER.

To enhance detection selectivity with a high speed of the compound identification in the presence of interference components a mobile chromato-mass-spectrometer (MCMS) is developed [2].
NIST/EPA data base containing up to 120 thousand compounds is used for mass-spectra interpretation. For parameters setting and results processing an external IBM PC is used. The detection time and sensitivity of the MCMS are much the same as of gas chromatographs.
The device is shown on Fig. 11, and its main characteristics are given below.

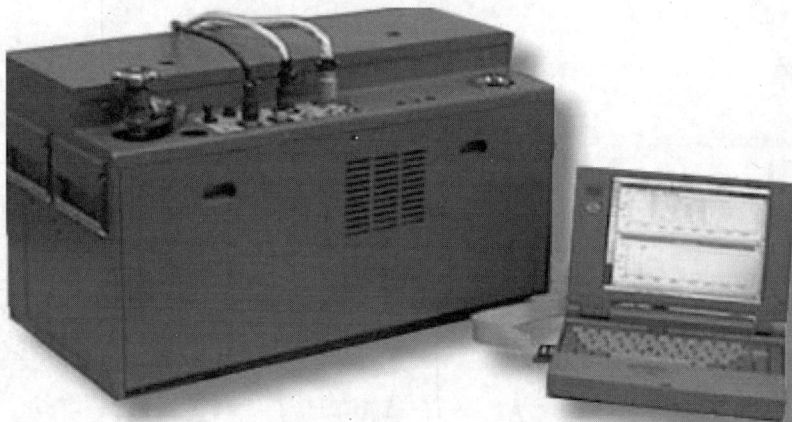

Fig.11: Mobile chromato-mass-spectrometer

FEATURES:

- Special-purpose chromatographic system for express compound detection in different media which performs 10-100 analysis per hour;
- Independent sampler for air sampling;
- Combined sample injector for analyzing gaseous and liquid samples;
- Continuous cyclic automatic air analysis;
- Electron impact ionization source;
- Mass-analyzer: magnetic, static with dual focusing;
- NIST/EPA data base and AMDIS program for mass-spectra interpretation;
- Software for automatic detection of given compounds in a sample;
- unique special-purpose compact vacuum system.

SPECIFICATIONS:

Detection limit in air:

- continuos cyclic automatic analysis - 10^{-5} mg/l
- enrichment mode (with sampling time of 10 s) 10^{-7} mg/l

Detection limit from extracts -	10^{-5} mg/l
Measurement interval -	10-100 analysis/h
Dimensions -	710x435x320 mm
Weight -	90 kg
Power consumption -	250 W
Mass range -	12-600 AMU
Maximum scan velocity -	3 spectra/s
Operating conditions according to -	1.7 GOST B20.39.304-76

3. ION MOBILITY INCREMENT SPECTROMETRY.

To enhance the speed of response of explosive detectors a technique of ion mobility increment spectrometry (IMIS) is being developed [3]. A possibility to detect TNT traces at a concentration of 10^{-14} g/cm^3 in several minutes without preconcentration is shown. The effect of the climatic conditions on the spectrometer response value is studied.

In summary it may be said that the above-mentioned techniques and devices can be used under field conditions to detect CW-agents, drugs and other toxic compounds.

REFERENCES

1. Gruznov V. M., Filonenko V.G. and Shishmarev A.T. (1999) J. of Analytical Chemistry, v.54, # 11, 1134.
2. Makas A. L., Troshkov M. L.(2001) Proceedings of the Seventh International Symposium on Protection against Chemical and Biological Warfare Agents, Stockholm, Sweden.
3. Buryakov I.A., Krylov E.V., Nazarov E.G.(1993), U.Kh. Rasulev. Int. J. of Mass Spect. and Ion Pros., 128, 143-148.

NOVEL TECHNIQUE FOR ULTRA SENSITIVE DETECTION OF ORGANIC COMPOUNDS

S.S.ALIMPIEV,*, S.M.NIKIFOROV,* A.A.GRECHNIKOV**, J.A.SUNNER***
*Prokhorov General Physic Institute Russian Academy of Sciences, Vavilov str.38, 119991, Moscow, Russia;

**Vernadsky Institute of Geochemistry and Analytical Chemistry Russian Academy of Sciences, Kosygin str.19, 117975,Moscow, Russia;

*** Montana State University, Bozeman, Montana, USA, 597171.

INTRODUCTION

State-of-art mass spectrometric methods are among the most successful analytical techniques for high sensitivity detection of organic compounds. However, there are also obvious problems. High-performance, commercial instruments are relatively complex and expensive and can be used only under laboratory conditions. Furthermore, sampling is generally complex and time-consuming. It has not yet been possible to adapt high-performance mass spectrometric methods to field conditions. For example, it is well recognized that a trained dog's nose has superior sensitivity for some compounds under field conditions. Adaption of mass spectrometric methods for field use and the development of portable instruments for ultra-sensitive detection and quantitative analysis of trace compounds in ambient air is an extremely important goal.

Time-of-flight mass spectrometers with ionization by pulsed laser hold great promise for field applications because of their relative simplicity and inherently high sensitivity for ion detection. However, present laser ionization methods are not suitable for organic analysis. Laser ionization of gaseous compounds is characterized by relatively low selectivity and strong fragmentation. Matrix-assisted laser desorption ionization (MALDI) has a much too high matrix chemical background in the low mass region and sample preparation is too complex and time-consuming for field use. Novel laser-induced ionization methods are therefore needed.

A few years ago, we introduced the use of suspensions of inorganic (carbon) or organic crystals in frozen analyte solutions in water or glycerol for laser desorption/ionization mass spectrometry [1,2]. Mass spectra were obtained of organic compounds, as well as of peptides and small proteins, by irradiating the suspensions with a 3.28-um IR or 337-nm UV laser pulses. The formation of ions was initiated by a charge separation of analyte ions from counter ions on the surface, followed by a heating-induced desorption of preformed (or pre-existing) ions. Because the surface structure was critical to obtaining mass spectra, the method was named "Surface-Assisted Laser Desorption Ionization" or SALDI.

M. Krausa and A. A. Reznev (eds.),
Vapour and Trace Detection of Explosives for Anti-Terrorism Purposes, 101-112.

GAS-PHASE SALDI

To develop practical SALDI ionization methods for detection of gas-phase compounds it was suggested to replace the suspensions of carbon or silicon in frozen water solutions by microscopically rough surface of carbon or silicon with adsorbed water [3]. A schematic of a SALDI instrument is shown in *Fig.1*.

The gas mixture to be analyzed is sampled to the vacuum chamber, where gas-phase analytes are adsorbed and ionized on the specially prepared solid surface. The preformed ions are desorbed with a pulsed laser, separated in a time-of-flight mass spectrometer (TOF) and detected in the ion detector. This gas-phase version of SALDI was rather successful and patent has been applied for [4].

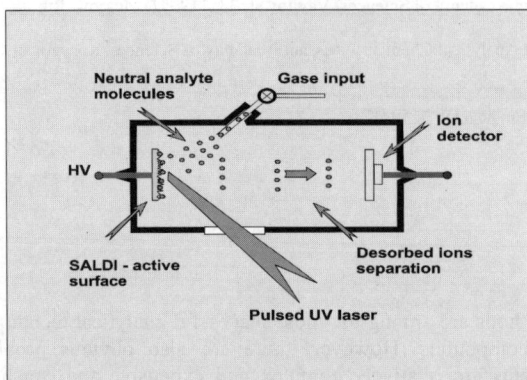

Fig.1. Schematic of gas-phase SALDI instrument.

CHARACTERISTIC FEATURES OF GAS-PHASE SALDI

Main characteristic features of SALDI are follows:

1. The main ions produced are protonated (positive ions) or deprotonated (negative ions) analyte molecules. The measurements were performed for both positive and negative ionization modes.

2. Mass spectra are "clean" with almost no chemical background and with little or no fragmentation of ions of analytes. This is due to the fact that ionization is very soft and no matrix is added. It is in contrast to some other well-known mass spectrometric techniques based on electron impact ionization or MALDI, in which the extensive fragmentation of organic molecules and matrix ions background occurs.

3. High ionization efficiency and excellent sensitivity for determination of such practically important compounds, such as narcotics, explosives, and rocket fuels.

4. Physical and chemical properties of the rough solid ionization surface are critical to the processes of adsorption, ionization and desorption.

5. The SALDI method may be realized in a comparatively simple, affordable, and highly efficient instruments for field-use. By having all ions start from a well-defined surface in a well-defined time, the ionization method is perfectly matched to TOF analyzers – the most simple and inexpensive mass spectrometers. So the full potential of the TOF analysis for a very high mass resolution may be realized. Also this method requires simple and accessible pulsed lasers.

PREPARATION OF IONIZATION SURFACE FOR GAS-PHASE SALDI

As has been shown in our previous investigations [3,4], the ionization efficiency of organic compounds on SALDI surfaces is strongly dependent on the characteristic size of surface structure irregularities. *Fig.2* illustrates the influence of surface structure on ion yield. From top to bottom, *Fig.2* shows mass spectra of diethylamine (m/z=73) obtained from porous silicon (characteristic size of structure irregularity ~100 nm), oxygen-atom-etched graphite (~500 nm) and fluorine-atom-etched silicon (more than 500 nm).

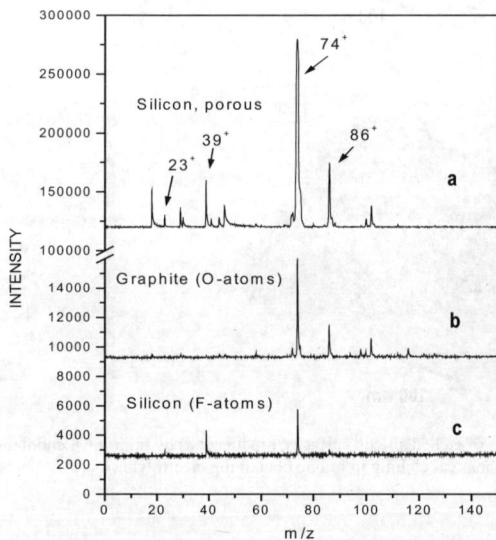

Fig.2: Comparison of mass spectra of diethylamine obtained from surfaces with different sizes of structure irregularities: a) – porous silicon, b) – O-etched graphite, c) – F-etched silicon.

It is seen, that the mass spectrum obtained from porous silicon is about 50 times more intense than the spectrum obtained from F-etched silicon. These results show that both the size of roughness irregularity and surface material are very significant for the ionization efficiency.

In order to further decrease the characteristic size of surface roughness, we modified the standard technique of obtain nano-sized structures on silicon surfaces [5]. This modification included electrochemical etching of silicon wafers in iodine-containing electrolytes followed by further modification of the etched surfaces by UV laser irradiation in the presence of water vapor.

Atomic force microscopy images of silicon surfaces, shown in *Fig.3*, demonstrate that the modification of the electrolyte composition resulted in a significant decrease of the characteristic size of surface structure irregularities as compared with the conventional anodizing technique.

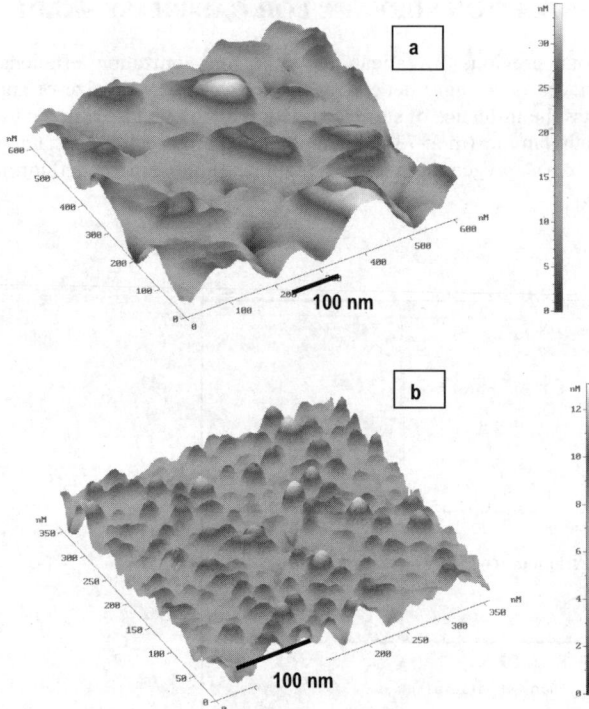

Fig.3: Atomic force microscopy images of etched silicon surfaces, produced: a) by traditional anodizing etching; b) by electrochemical etching in iodine containing electrolyte.

A comparison of ionization efficiencies of silicon surfaces produced by traditional anodizing etching and by electrochemical etching in iodine-containing electrolyte, respectively, is shown in *Fig. 4*. The dependencies of the signal intensity of protonated pyridine on the N_2-laser (337nm, 4 ns) fluence shows that modification of the etching technique dramatically increases the ionization efficiency and strongly decreases the laser fluence threshold.

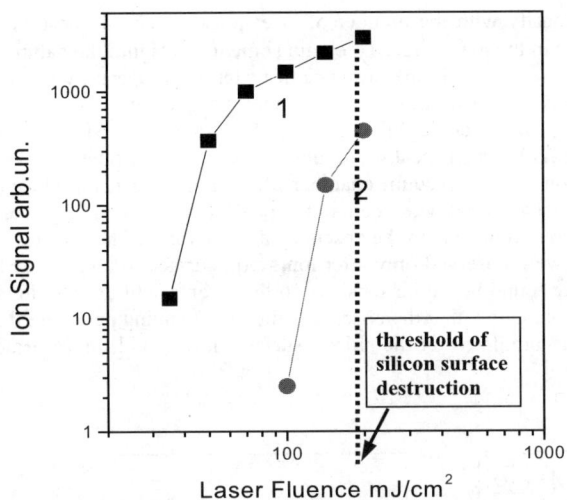

Fig.4: The intensity of the signal of protonated pyridine obtained from silicon surfaces as a function of the laser fluence, produced by (1) traditional anodizing etching and (2) by electrochemical etching in iodine containing electrolyte.

A further increase of the surface ionization efficiency on etched silicon substrates was observed with the number of UV laser pulses used to irradiate the same spot on the surface. This effect is illustrated in *Fig.5*. It is seen that there was a strong enhancement of the ion yield from freshly prepared silicon surface during the first 10^3 laser pulses and that the signal stabilized after irradiation by $\sim 10^4$ pulses.

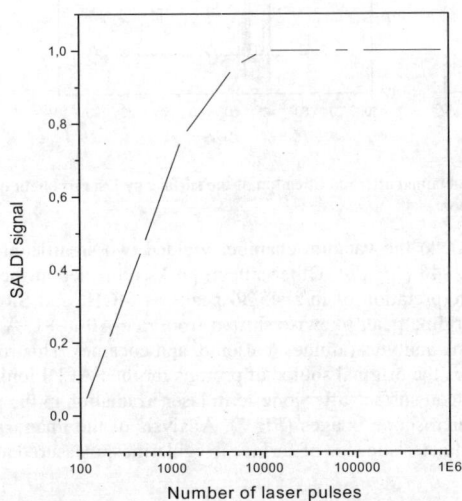

Fig.5: Dependence of the SALDI signal for protonated pyridine from etched silicon on the number of N_2 - laser pulses used to irradiate the same spot on the surface. The water vapor was $P=10^{-5}$ Torr, and the laser fluence on the surface was 50 mJ/cm^2.

The increase in the signal intensity with the number of laser pulses was only observed when water vapor ($\approx 10^{-5}$ Torr) was present in the ion source. To obtain some insights into the nature of the surface modifications induced by the laser pulses in the presence of water, mass spectra were obtained on the silicon surfaces themselves. A typical spectrum, obtained under "wet" conditions, is shown in *Fig. 6a*. Prominent silicon-containing peaks include Si^+ (m/z=28), Si_2^+ (m/z=56), Si_3^+ (m/z=84), $(SiOH)^+$ (m/z=45), and $(SiOSi)^+$ (m/z=72). Other peaks are due to K^+ and pyridine $(Pyr+H)^+$. (Pyridine remained at trace concentrations in the vacuum chamber after having been introduced into the ion source during preceding work.) The two oxygen-containing peaks (m/z=45 and 72) were absent in the spectra obtained under "dry" conditions, i.e. in the absence of water vapor. These peaks, as well as the silicon dimer and trimer ions, were observed only after long-term surface irradiation by the laser. The amplitudes of these peaks were found not to be sensitive to the laser repetition rate, in contrast to the protonated pyridine peak. This supports that the observed silicon-containing peaks are due to products of the surface modifications accumulated on the silicon surface during multi pulse irradiation in the presence of water vapor.

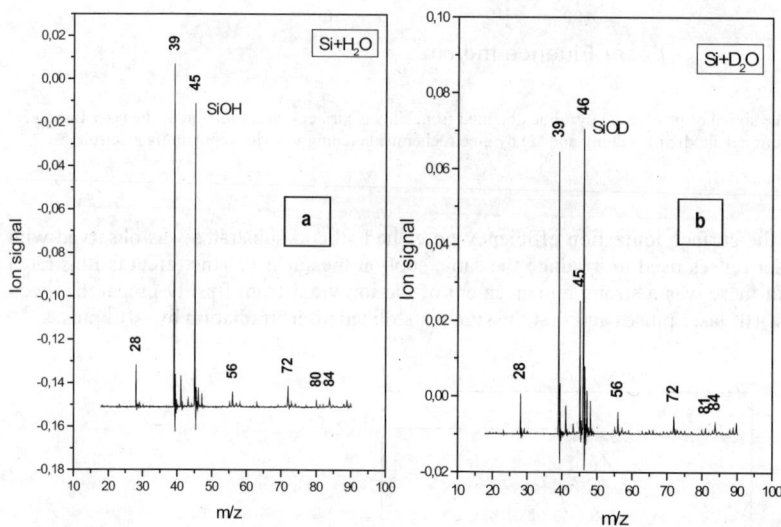

Fig.6: Mass spectra of etched silicon surfaces, obtained after modification of the surface by laser irradiation in the presence of (a) H_2O water vapor and (b) D_2O water vapor.

The addition of heavy water vapor, D_2O, to the vacuum chamber yielded two significant results. The first was the shift of the m/z=45 peak to m/z=46 (*Fig.6b*). Other silicon peaks remained at the same masses, as expected. This strongly supports the interpretation of m/z=45/46 peaks as $SiOH^+$ and $SiOD^+$, respectively. The second observation was that the pyridine peak likewise shifted from m/z=80 to 81. A similar one mass unit shift was observed for all tested basic analytes (amines, caffeine, and cocaine). This result shows that it is water inside the vacuum chamber that is the original source of protons for the SALDI ionization process.

The change in the morphology of the silicon surface after long-term laser irradiation in the presence of water vapor is also clearly seen in electron microscope images (*Fig.7*). Analysis of the images revealed that the laser irradiation did not change the characteristic size of surface irregularities, measured along the plane of the silicon substratum, but strongly increased image contrast. Our observations also showed that the initial surface roughness plays a key role for the surface modification. Thus, the strong increase in the ionization efficiency with continuing laser irradiation, seen in *Fig. 5,* has been observed also for silicon surfaces made rough by different techniques (ion bombardment, sub-micron surface grinding (sanding) by diamond powders). In contrast, no analyte signal was obtained from initially polished silicon surfaces, even after long term laser irradiation in the presence of water vapor.

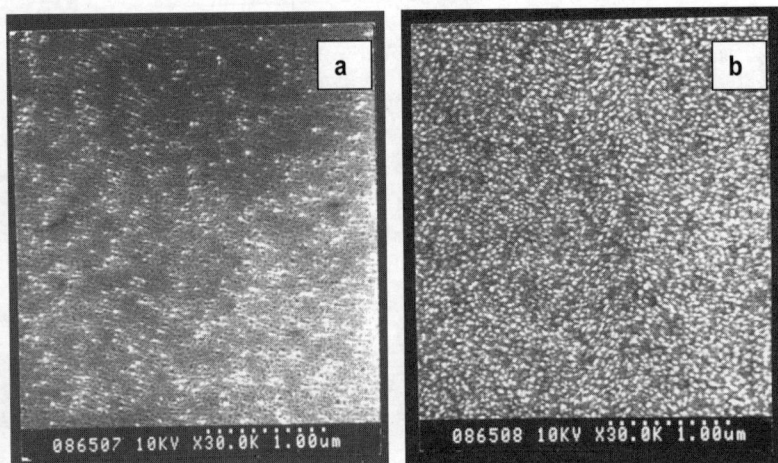

Fig.7: Electron microscope images of silicon surface (a) before modification by laser irradiation in the presence of water vapor and (b) after modification by 10^4 laser pulses.

Fig. 8 shows mass spectra of caffeine (protonated molecule at m/z=195 with major fragment at m/z=138) obtained with and without water vapor addition. The effect of increasing the ionization efficiency by the addition of water vapor to the vacuum chamber was observed in our early investigations [3]. It was shown that this increase was most prominent for analytes with relatively low aqueous basicities (i.e. low pKa values for the conjugate acids).

The results obtained for gas-phase SALDI, illustrated in *Fig. 4-8* can be interpreted by a model of laser assisted etching of rough silicon surfaces in the presence of water vapor. Such etching seems to be a multistage process. The first stage is a chemical activation of the surface. After electrochemical etching, the silicon surface is passivated by hydrogen, fluorine and/or iodine. Laser irradiation leads to desorption of this passivated layer and the "activation" of silicon surface bonds. The second stage is adsorption of water on the activated surface. In the third stage, adsorbed water molecules dissociate on the surface. Such dissociation requires two adjacent dangling bonds per water molecule, leading to OH and H fragments on each dangling bond. Some hydroxyl groups could dissociate to form a Si-O-Si species on the surface, as observed in the mass spectra (*Fig.6*). The observation of silicon dimers and trimers from the laser-modified silicon surfaces could be explained by a weakening of silicon-silicon crystal bonds after surface Si atoms have formed bonds with electronegative hydroxyl groups, leading to the desorption of silicon clusters at a relatively low laser fluence. An alternative explanation could involve diffusion and association of interstice silicon atoms on the tips of surface structures and the desorption of such clustered silicon atoms. But this raises the question of why this process is facilitated by the presence of water, and this needs further investigations.

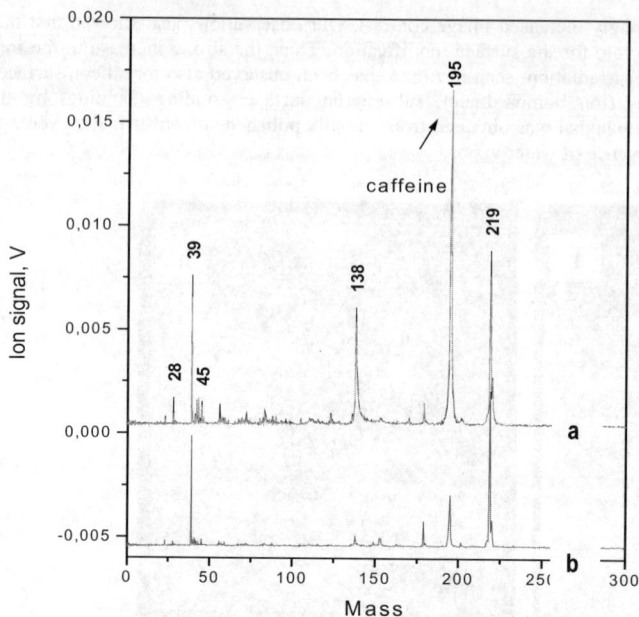

Fig.8: Mass spectra of caffeine obtained (a) with and (b) without water vapor addition.

According to this model, adsorbed water plays the key role in the process of ionization of aliphatic amines, hydrazine, narcotics and other organic compounds with high aqueous basicities. When water is dissociatively adsorbed, the surface becomes covered by hydroxyl groups. Neutral analyte molecules M adsorbs on the surface and form complexes with such groups. Proton transfer in Si-O···H-M would result in the formation of protonated molecules (H-M)$^+$ which are desorbed by laser irradiation. According to this mechanism, only sufficiently basic molecules will be protonated. As previously reported [3], for effective ionization, the aqueous pK$_a$ value of the protonated analyte must be higher than about 4.

As have been shown in *Fig. 3 and 4,* the ionization efficiency strongly increases with decreasing size of the surface roughness. This observation could be explained by generation of relatively strong electrostatic fields between the "peaks" and "valleys" on the rough surface, during and after laser irradiation. Interstice silicon atoms are accumulated in the protruding "peaks" whereas vacancies are localized in the valleys. In such case, carrier generation in laser field or even surface heating will provide positively charged tips and hence electrostatic fields which strength will obviously increase with decreasing characteristic size of surface irregularities. Such electric fields might strongly facilitate the charge separation in Si-O···H-M complexes and the formation and desorption of (H-M)$^+$ analyte ions.

EXPLOSIVE DETECTION BY GAS-PHASE SALDI

The results reported above were obtained with basic compounds, which are detected as protonated molecules. Protonation is applicable to many important classes of compounds, such as narcotics, drugs, rocket fuels, amino acids and proteins. However several compound classes, many of which are environmentally important, are best observed in negative ion mode. These include explosives such as nitro-derivatives of toluene. The gas-phase SALDI mass spectrum of TNT is presented in *Fig.9*. It is seen that the base peak is due to deprotonated molecules (TNT-H)$^-$ (m/z=226). Other minor peaks are due to the molecular ion (TNT)$^-$ (m/z=227) and two fragments: (TNT-O)$^-$ (m/z=210), (TNT-2O)$^-$ (m/z=194).

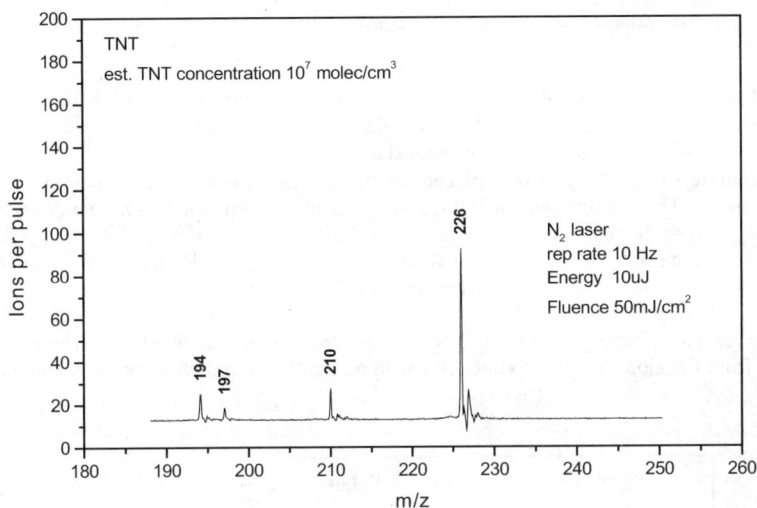

Fig.9: Mass spectra of trinitrotoluene.

The main analytical characteristics of SALDI detection of TNT have been studied in more details in special investigations.

LIMIT OF DETECTION

The estimation of the limit of detection of SALDI is a rather difficult task. This is due to the different reasons: very high sensitivity, low saturated vapor pressures of analyzed compounds, active surface of the vacuum chamber and others. In order to solve this task, it was necessary to create the calibrated microflow source of analytes. We achieved by using a specially designed bulk acoustic wave resonator, shown in *Fig.10*.

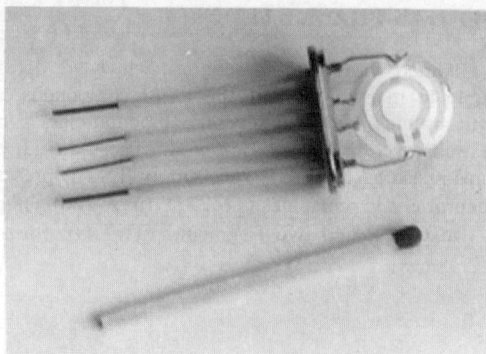

Fig.10: Bulk acoustic wave resonator as a source of microflows.

The main characteristics of the resonator are as follows: the operating frequency is 16.5 MHz, the Q-factor is more than $5 \cdot 10^4$, and the outer layer of the electrodes is gold. The built-in microheater gives the opportunity to work at different temperatures on the surface.

A small amount of analyte (about 20 μg) was placed on the outside surface of an electrode. The exposed surface area of the TNT sample was about 20 mm². It is well known that a change in the mass of a thin coating on an acoustic wave resonator corresponds to a change of resonance frequency. Thus, it is simple to determine the rate of loss of analytes by measuring the change of the resonator frequency with time. Such dependencies for trinitrotoluene (TNT), obtained at different pressures of ambient air are presented in *Fig 11*.

Fig. 11 shows that the resonator frequency is linear over a wide range of sublimation times. The rate of TNT loss is obtained from the slope and the frequencies with no loading and with a known loading on TNT.

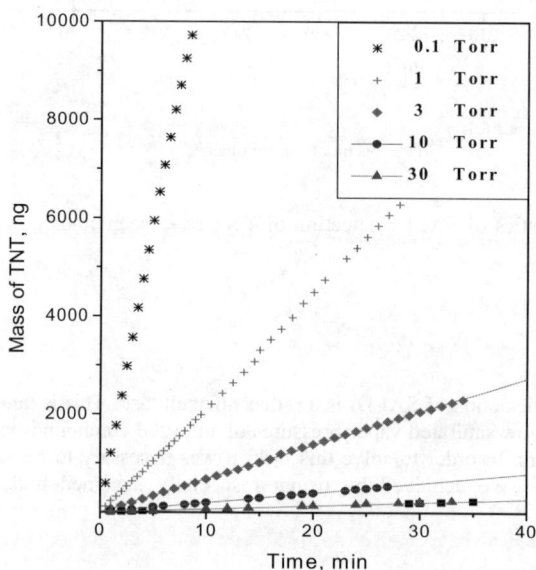

Fig.11: The change of resonator frequency with time. The increase in the frequency reflects the gradual loss of TNT at various pressures of ambient air.

Fig. 12 shows the measured rates of TNT sublimation as a function of ambient air pressure. One can see that the rate of sublimation of TNT was about $3 \cdot 10^{14}$ molecules/s in vacuum and decreased to $6 \cdot 10^{10}$ at atmospheric pressure. The obvious reason of this decrease is the slow diffusion of TNT molecules into air. The measured rate of sublimation of TNT into vacuum is much too high for the high sensitivity of SALDI. To decrease and regulate the rate of emission of analyte molecules in vacuum, we mixed the TNT analyte with an inert membrane polymer (usually a glassy polymer with a large free volume). This procedure made it possible to create a stable analyte flow in vacuum of 10^{10} molecules/s and less.

With the use of measured molecular emission rates, we estimated limits of detection of SALDI for various active surfaces. The best results were obtained for the etched silicon substrates shown in *Fig.3* (b). At the limit of detection, the flow of nitro-derivatives of toluene on the active surface was estimated to be $5 \cdot 10^9$ molecules/cm$^2 \cdot$s, which corresponds to an analyte concentration of 10^6 cm^{-3}. In positive ion mode, for aliphatic amines and narcotics, the sensitivity was higher and the detectable analyte concentration was estimated to be $5 \cdot 10^4$ cm^{-3}.

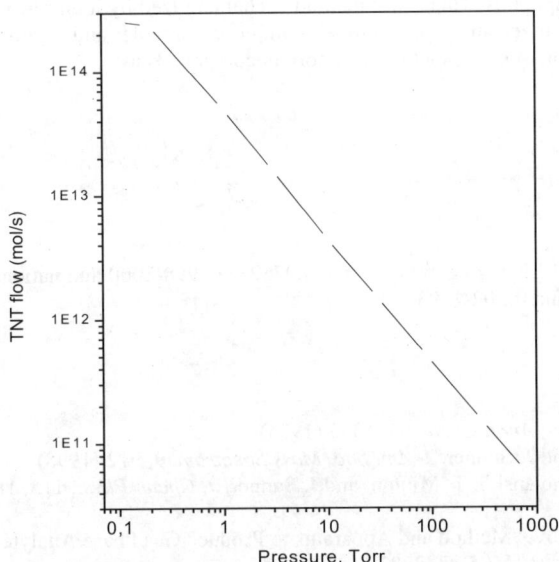

Fig.12: The dependence of the rate of sublimation of TNT on the ambient pressure.

SELECTIVITY

The lack of selectivity is one of the main limiting factors in the use of laser mass spectrometric method for on-line analysis, but the SALDI method is a fortunate exception. The high selectivity of SALDI is caused by several factors. First of all, it is due to the selectivity of the ionization process. As mentioned above, positive mode ionization in SALDI is based on protonation of analyte molecules, which is strongly selective for basic analytes. In particular, no background signal is obtained for the main constituents in ambient air, such as nitrogen, oxygen, water, etc. This represents a very significant advantage for SALDI also over electron ionization, the most commonly used ionization method in ambient air detectors.

Additional selectivity of the SALDI process is due to the fact that it is a very soft ionization process with limited but very specific fragmentation of analyte ions. For example, the SALDI mass spectrum of caffeine shows only two peaks, at $m/z=195$ and 138, the mass spectrum of cocaine similarly has two peaks at $m/z=304$ and 182, and the negative ion mode mass spectrum of TNT shows four ion peaks (see *Fig.8*). The simultaneous monitoring of both a major fragment ion peak and the molecule ion peak dramatically increases the ability to discriminate a targeted analyte from interfering chemical noise.

Finally, a relatively high mass spectrometric selectivity is obtained as a relatively high resolution, $m/\Delta m$, of ~1000 is easily realized in a simple linear TOF MS due to short desorbing laser pulse and to a well defined ionization zone, located on the active SALDI surface.

CONCLUSIONS

Surface-Assisted Laser Desorption Ionization (SALDI) has been developed for the ultra sensitive detection of gas-phase organic compounds. This novel method is characterized by high sensitivity and selectivity. The method may be realized in a comparatively simple, affordable, and highly efficient instrument for on-line trace detection of such practically important compounds as:
- narcotics;
- biomolecules;
- explosives.

ACKNOWLEDGMENT

This work was supported by NATO Linkage grant (PST.CLG.975233) 1998-2000 and national RFBR grants 01-02-16617; 02-02-08114 and 02-02-17936.

REFERENCES

1. J. Sunner, E. Dratz, and Y.C. Chen, *Anal. Chem.* 67, 4335 (1995).
2. P. Kraft, S. Alimpiev, E. Dratz, and J. Sunner, *J. Am. Soc. Mass Spectrom.* 9, 912 (1998).
3 S. Alimpiev, S. Nikiforov, V. Karavanskii, T. Minton, and J. Sunner, *J. Chem. Phys.*, 115, 1891 (2001).
4. J. Sunner, S. Alimpiev, S. Nikiforov, "Method and Apparatus to Produce Gas Phase Analyte Ions", *USA Patent application #US 2002/0121595 A1.* 2002.
5. S. Alimpiev, S. Nikiforov, A. Grechnikov, V. Karavanskii, and J. Sunner, "Method of forming of rough surface of silicon substrates and electrolyte for anode etching of silicon substrate", *RF Patent application #RF 2003/101425.* 2003.

THE ANALYTICAL CHARACTERISTICS OF ION MOBILITY INCREMENT SPECTROMETER DURING THE DETECTION OF EXPLOSIVE VAPOURS AND PRODUCTS OF THEIR DEGRADATION

IGOR BURYAKOV

The Design & Technological Institute of Instrument Engineering for Geophysics and Ecology (IDE), the Siberian Branch of RAS, 3/6 Pr. Ak. Koptyuga, 630090, Novosibirsk, Russia, E-mail: buryakov@uiggm.nsc.ru

ABSTRACT:

The operational detection of trace quantities of explosive vapors and products of their degradation (EV) in air is a complex analytical problem. High demands placed on device sensitivity and selectivity are dictated by a rather low vapor pressure of the compounds, their high adsorption power and interfering components present in air in large quantities. Ion mobility increment spectrometer (IMIS) is one of the instruments of those satisfying these requirements.

Like an ion mobility spectrometer (IMS), the operation of IMIS rests on sampling air containing a mixture of trace constituents, its ionization, spatial separation of produced ions and separated ions detection. IMIS differs from IMS in that ions of different types are separated in IMIS by ion mobility increment that depends on electric field strength.

In this work we investigated the possibility of the selective registration of vapors of 2,4-dinitrotoluene, 2,4,6-trinitrotoluene, pentaerythritol tetranitrate 1,3-dinitrobenzene, 1,3,5-trinitrobenzene with the aid of IMIS with different humidity of air, in the presence of vapors of gasoline, diesel fuels, ammonia. The calculated detection limit, sensitivity, linearity and speed of response of IMIS on detecting vapours of the above-mentioned compounds have been determined.

KEY WORDS:

Ion mobility increment spectrometer, explosives.

M. Krausa and A. A. Reznev (eds.),
Vapour and Trace Detection of Explosives for Anti-Terrorism Purposes, 113-121.
© 2004 *Kluwer Academic Publishers. Printed in the Netherlands.*

INTRODUCTION

The expeditious detection of trace quantities of explosive vapors and products of their degradation in air in real time is a complex analytical problem. High demands placed on device sensitivity and specificity are dictated by a rather low vapor pressure of the compounds, their high adsorption power and interfering components present in air in large quantities. One of the devices that meets the requirements is an ion mobility increment spectrometer (IMIS) [1, 2]. The names like ion non-linear drift spectrometer, high-field asymmetric waveform ion mobility spectrometer, field ion spectrometer, radio-frequency-based ion-mobility analyzer are used as a synonym for IMIS.

Like an ion mobility spectrometer (IMS), the operation of IMIS rests on sampling air containing a mixture of trace constituents, its ionization, spatial separation of produced ions and separated ions detection. IMIS differs from IMS in that ions of different types are separated in IMIS by ion mobility increment that depends on electric field strength [3-5].

ION SEPARATION IN IMIS

Ion drift velocity, V, caused by an action of electric field (E) is [6]:

$$V = K(E)E; \quad K(E/N) = K_0 \left(1 + \alpha(E/N)\right) = K_0 \left(1 + \sum_{n=1}^{\infty} \alpha_{2n} \left(\frac{E}{N}\right)^{2n}\right) \quad (1)$$

where K_0 $(cm^2(V \cdot c)^{-1})$ is the mobility coefficient in a weak field (E/N < 6 «townsends», or Td, where 1 Td = 10^{-17} V·cm²), N is a neutral gas density, α(E/N) is a normalized function which describes the electric field dependence of the mobility – the mobility coefficient increment, α_{2n} are decomposition coefficients, $n \geq 1$ is an integer. Under the action of periodic alternating asymmetric waveform field, $E_d(t) = E_d \cdot f(t)$, (E_d is an amplitude, f(t) is a form of the field) that meets the conditions [3]:

$$\int_t^{t+T} f(t)dt = 0, \quad \frac{1}{T} \int_t^{t+T} f^{2n+1}(t)dt \equiv <f^{2n+1}> \neq 0, \quad (2)$$

ions executing fast oscillatory motions with period, T, drift with velocity $\langle V_i \rangle$ characteristic for each i-th component proportional to $\alpha_i(E/N)$ (broken brackets mean averaging over a period). Should the ion drift of the i-th component be compensated by constant electric field, E_{ci}, the ion velocity will be zero :

$$\langle V_i \rangle = \langle K_0(1+\alpha_i)(E_d f(t) - E_{ci}) \rangle = 0 . \quad (3)$$

The compensation field of the i-th component is expressed by substituting decomposition of coefficient α from (1) into Eq. (3) and using condition (2) and $(E_d - E_c)^n \approx E_d^n - nE_d^{n-1}E_c$, $|E_d| \gg |E_c|$ [4]:

$$E_{ci} \approx \frac{E_d \sum_{n=1}^{\infty} \alpha_{i2n} \left(\frac{E_d}{N}\right)^{2n} <f^{2n+1}>}{1 + \sum_{n=1}^{\infty} (2n+1)\alpha_{i2n} \left(\frac{E_d}{N}\right)^{2n} <f^{2n}>} \tag{4}$$

The ions for which $\langle V_i \rangle = 0$ are transported to a collector with a carrier-gas, the other ions leave a separation area and recombine. With E_c changing, spectrum of a mixture of ions of all types is recorded.

In this work we studied the possibility for the selective vapor detection of: 2,4-dinitrotoluene, 2,4,6-trinitrotoluene, pentaerythritol tetranitrate, 1,3-dinitrobenzene, 1,3,5-trinitrobenzene with the aid of IMIS with different air humidity, in the presence of gasoline, diesel fuels and ammonia vapors. The calculated detection limit, linearity and speed of response of IMIS on detecting vapours of the above-mentioned compounds have been determined.

EXPERIMENTAL

Apparatus

Block-diagram of IMIS is given on *Fig. 1*. IMIS comprises a heated ionization chamber (sampling flow rate $Q_s = 0.5 \div 5$ cm^3/s, β—source ^{63}Ni, temperature $t_i = 50 \div 150°C$), an ion separation chamber formed by two electrodes which are coaxial cylinders, the chamber is purged with a carrier gas (dry air with a water vapor concentration ≤ 100 ppm, flow rate, $Q_{tr} = 50$ cm^3/s), gas seal, for transporting ions from the ionization chamber into the separation chamber, an ion collector, an electrometer, a compensation voltage generator, filtration system. To set up field in the separation chamber equal to:

$$E_d(t) = U_d \times f(t)/d, \tag{5}$$

a dispersion voltage generator was used. The generator was connected to the electrodes of the separation chamber and had the following parameters:
voltage form - is a function normalized per unit which describes field waveform (*Fig.1*):

$$f(t) = (\sin[\pi \cdot (t-mT) / \tau] - 2\tau / \pi T) / (1 - 2\tau / \pi T), \quad \text{with } mT \leq t \leq (mT+\tau);$$
$$f(t) = - (2\tau / \pi T) / (1 - 2\tau / \pi T), \quad \text{with } (mT+\tau) \leq t \leq (m+1)T,$$

where m $\geq$ 0 – is an integer. High-voltage pulse duration τ = 1.9 μs; period T= 5.9μs; dispersion voltage amplitude U_d = 1÷4 κV. To set up compensation field E_c in the separation chamber the electrode of a small diameter was connected to voltage source U_c.

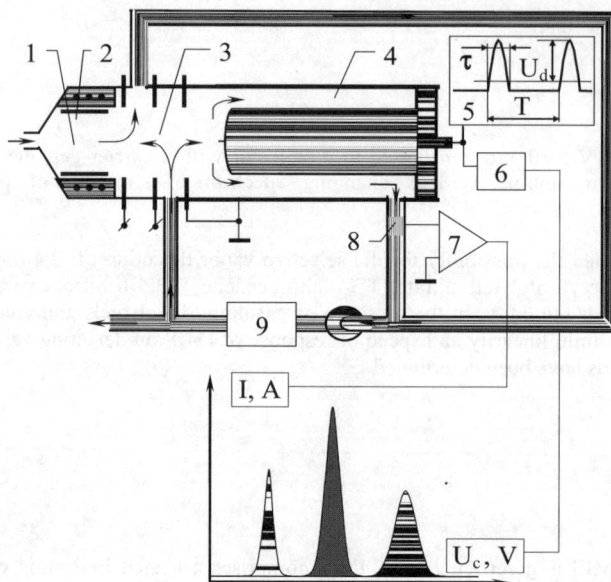

Fig. 1. Block-diagram of IMIS : 1 – ionization chamber, 2 – β–source ^{63}Ni, 3 – gas seal, 4 – separation chamber, 5 – dispersion voltage generator, 6 – compensation voltage generator, 7 – electrometer, 8 – ion collector, 9 – filtration system.

Reagents

Samples of 1,3–dinitrobenzene (DNB), 1,3,5– trinitrobenzene (TNB), 2,4–dinitrotoluene (DNT) and 2,4,6–trinitrotoluene (TNT), pentaerythritol tetranitrate (PETN) of the highest purity were obtained from the Research Institute of Special Technics and Communication (Novosibirsk, Russia). To obtain air-vapor mixtures of nitrocompounds, purified air flow was passed through a quartz tube (∅ 0.3 cm, 20 cm in length) the inside of which was covered with a tested compound. The air flow saturated with explosive vapors was mixed with the input flow. By varying the input flow we changed the nitrocompound concentration. The concentration of gasoline, diesel fuels or ammonia was 1.5, 2, 0.02 mg/l.

IONIZTATION PROCESSES

Tab.1 shows the main types of ion-molecule reactions into which the molecules of tested compounds enter being ionized by β–source at atmospheric pressure, mass-charge ratio (m/z) and types of ions produced (M is a compound molecule, H is hydrogen atom). Since the molecules of DNB and TNB have a strong electron affinity, ions of these compounds are produced by the reactions of associative electron capture with the formation of M^- ions [7-9]. The reaction of proton abstraction yields DNT and TNT ions, and the basic reaction products are $(M–H)^-$ ions. This is explained by the fact that DNT and TNT are gas-phase acids, and electrons are pulled to substituting groups NO_2 of a benzene ring, which causes hydrogen of a methyl group to acquire acidic properties [10-11]. PETN ions are produced by the reactions of electron transfer and electrophilic capture with the formation of M^- or $M*NO_3^-$ [12].

Tab. 1: The main types of ion-molecule reactions and ions.

Compound	Reaction	m/z	Ion
DNB	associative electron capture	168	M^-
TNB		213	
DNT	proton abstraction	181	$(M–H)^-$
TNT		226	
PETN	Electrophilic capture	316	M^-
		378	$M*NO_3^-$

SELECTIVE DETECTION OF NITROCOMPOUND VAPORS IN AIR WITH IMIS

The IMIS spectrum is the compensation voltage dependence of the ion current, $I(U_c)$, each ion type is detected when Eq.4 holds and are dependent on individual ion parameters (mass, structure, interaction potential), dispersion voltage amplitude, U_d, atmospheric pressure and drift gas composition. On *Fig. 2* drift spectra of air containing vapours of : a) DNT, TNT, PETN; b) DNB, TNB under negative mode are given. It is evident from the *Fig.2* that the ion peaks of each compound have compensation voltage values, U_c, typical of a particular compound. The fact that the peaks belonged to the tested compounds was determined with a gas chromatograph (GC) [13]. An agreement between compensation voltages, U_c, of the peaks detected by IMIS on analyzing vapour phase and chromatographic fractions confirms indirectly that the peaks belong to the tested compounds.

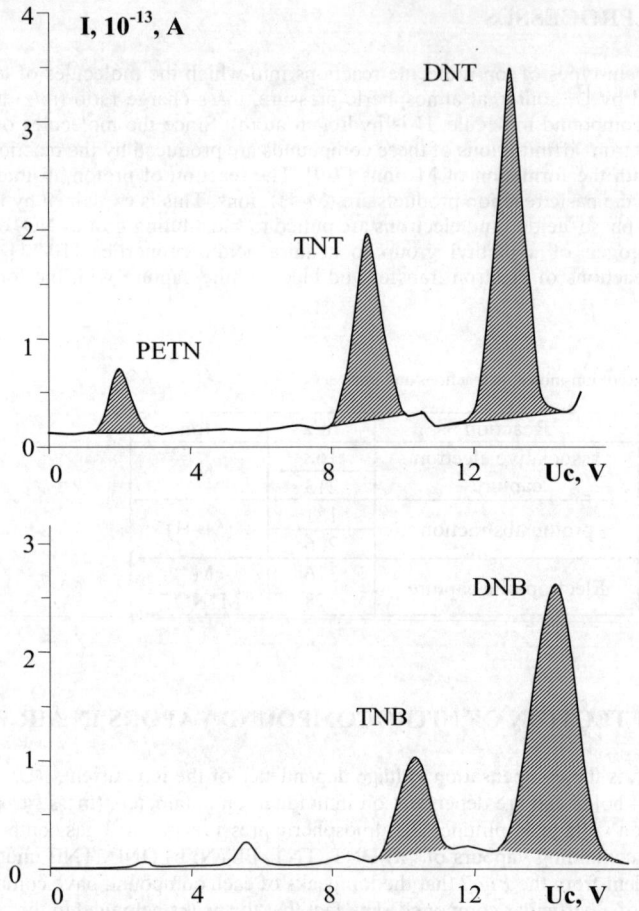

Fig. 2: Drift-spectra of air containing vapour of: a) DNT, TNT and PETN; b) DNB and TNB, obtained under negative mode.

ANALYTICAL CHARACTERISTICS OF IMIS

The curves of the ion current of IMIS versus DNB, DNT, TNB, TNT and PETN concentration are given on *Fig. 3* at the logarithmic scale. Since the data on the vapour pressure of these compounds given in literature differ, the X-scale is given in the units of concentration assigned to saturated vapour concentration (C/C_{sat}). Solid points indicate experimental signal value, and lines extrapolate the concentration dependence of the ion current intensity I(C) to a low value range. The intersection points of the extrapolation line and the line of double noise amplitude equal to $2 N_{oise} = 6 \times 10^{-15}$ Å are determined as values of a minimum detection limit (MDL). These values are calculated for DNB, DNT, TNB and TNT. The MDL of PETN was determined directly from the experiments. *Tab. 2* presents the values of MDL for DNB, DNT, TNB, TNT, PETN given in relative units (C/C_{sat}), linear dynamic range (LDR).

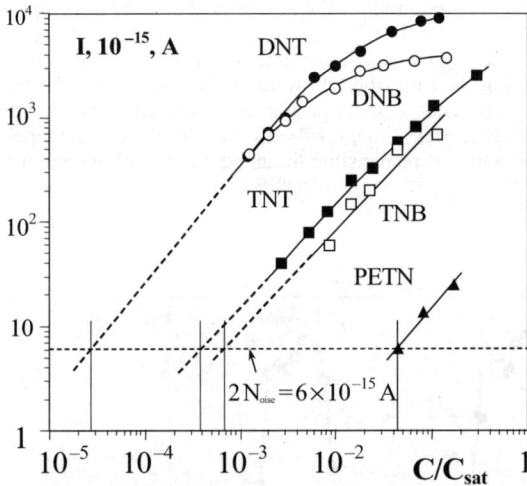

Fig. 3: Curves of ion current intensity detected by IMIS versus concentration of DNB, DNT, TNB, TNT, PETN expressed in relative units.

Tab. 2: Analytical characteristics of IMIS.

Compound	MDL, C/C_{sat}	LDR
DNB	2.7×10^{-5}	200
TNB	2.7×10^{-5}	400
DNT	7×10^{-4}	200
TNT	3.8×10^{-4}	250
PETN	4.3×10^{-2}	5*

* - range covered.

SPEED OF RESPONSE

Speed of response, t_R, is determined as the time required for the device to obtain 0,9 of complete signal variation amplitude. In IMIS, t_R, is determined by time, t_t, required to transport ions to the collector and electrometer speed, t_e. In theory t_t value is given by the volumes of the input lines, ionization and separation chambers, the values of sampling and transport flow rates (Q_{in}, Q_t), and for the device used $t_t \approx 0.6$ s, $t_e \approx 0.2$ s. During the experiments on determining IMIS time response, t_R, to TNT, a value of about 2 s was obtained *(Fig. 4)*. Aftereffect time, $t_a = 4 \div 5$ s. Such high t_R value appears to be due to sorption of TNT by the input lines, which is responsible for an additional delay upon the sample delivery to the ionization chamber.

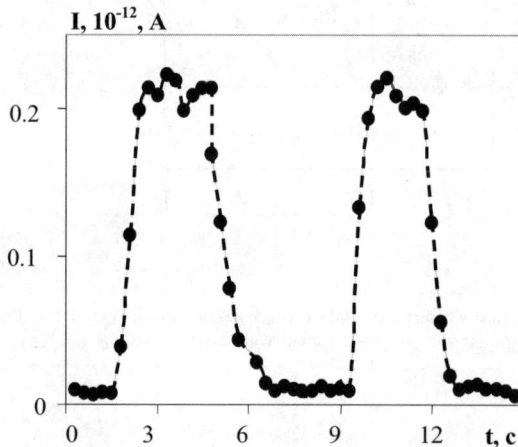

Fig. 4: IMIS response to varying concentration of TNT (300 ppt) in air.

SPECIFICITY

The specificity of an analytical method is characterized by its capability of detecting an analyte in the presence of interferents. To assess the specificity, known substances that can interfere with the determination of the analyte are introduced into the test sample. The simultaneous elution of the analyte and interferent can deteriorate the precision of the results and the limit of detection.

To determine the IMIS specificity we added water vapors to the input flow varying relative humidity. An increase in the relative humidity of the input flow from 35 % to 95% led to a reduction in the amplitude of the peaks : TNB, TNT, PETN – 1.2÷1.4 times; DNB, DNT – 5 times.

In this work investigated the influence of gasoline, diesel fuels and ammonia vapors in the concentrations 1.5, 2 or 0.02 mg/l, respectively. The presence gasoline or diesel fuel vapors in the input flow in the above-indicated concentrations had no a noticeable effect on the peak amplitudes of the substances being investigated, but the presence of ammonia led to reduction in the peak amplitude several times.

CONCLUSION

The results of testing have shown a possibility for selective detection of samples containing 1,3–dinitrobenzene, 1,3,5–trinitrobenzene, 2,4–dinitrotoluene and 2,4,6–trinitrotoluene, pentaerythritol tetranitrate with IMIS. Calculated detection limit, linearity and speed of response of IMIS on detecting explosives vapour are determined. The results of experiments point to high sensitivity, specificity and speed of the IMIS response. These properties of IMIS render it as the most promising device for expeditious detection of explosives.

REFERENCES

1. B. Carnahan, S. Day, V. Kouznetsov, A. Tarassov.// Proc. of 4th Int. Workshop on IMS, Cambridge, UK (1995), pp. 789-800.
2. Buryakov I. A., Kolomiets Yu. N., Luppu V.B. Detection of Explosive Vapors in the Air Using an Ion Drift Nonlinearity Spectrometer // Jour. of Anal. Chem., 2001, V. 56, No. 4, pp. 336-340.
3. Gorshkov M.P. // Pat. №966583 USSR. 1982.
4. Buryakov I.A., Krylov E.V., Nazarov T.G., Rasulev U.Kh. //Int. Jour. of Mass Spectrom. and Ion Processes. 1993. V. 128. pp. 143-148.
5. Buryakov I.A // Int. J. for Ion Mobil. Spect., 2001, V. 4, No. 2, pp. 112-116.
6. Mason E.A., McDaniel E.W. Transport Properties of Ions in Gas. New York: Wiley, 1988, p. 141.
7. P. Kebarle, S. Chowdhury, Chem. Rev., 87 (1987) 531.
8. K.A. Daum, D.A. Atkinson, R.G. Ewing, Int. J. for IMS, 4 (2001) 179.
9. G.R. Asbury, J. Klasmeier, H.H. Hill, Talanta, 50 (2000) 1291.
10. G.E. Spangler, J.P. Carrico, D.N. Campbell, J. of Testing and Evalution, JTEVA, 13 (1985) 234.
11. K.A. Daum, D.A. Atkinson, R.G. Ewing, W.B. Knighton, E.P. Grimsrud, Talanta, 54 (2001) 299.
12. Danylewych-May L.L. // Proceedings of the First Int. Symp. on Explosive Detection Technology, Atlantic City. 1992. P. 672.
13. Buryakov I.A., Kolomiets Y.N., Louppou V.B. // Int. J. for Ion Mobil. Spect., 2001, V. 4, No. 1, pp. 13-15.

POTENTIALS AND REQUIREMENTS OF AN ELECTRONIC NOSE FOR USE AS A VAPOR-BASED DETECTOR OF EXPLOSIVE PACKAGES

J. GOSCHNICK, T. SCHNEIDER,

Forschungszentrum Karlsruhe, Institut für Instrumentelle Analytik
Postfach 3640, D-76021 Karlsruhe

MOTIVATION

Even if bombs are hidden in unsuspicious objects, explosives packages release a characteristic gas ensemble which can be detected by sniffer dogs. In view of several draw-backs of sniffer dogs (such as limited working time per day, constrained age span they can be employed, high costs involved when employing a group of dogs) artificial olfaction with an Electronic Nose (EN) may present an interesting alternative. An EN can be applied 24h per day, has lower investment and operational costs and the spectrum of detectable gases is broader. However, the dog's nose is known to show an extreme sensitivity and gas discrimination power.

Although the vapor pressure of the explosive agent itself is often very low at room temperature (vppt range for TNT, s. *Tab. 1*), the necessary periphery to make up the complete explosive package frequently releases much more vapor into the ambient air. Its impurities, additives, detonator and package materials have much higher vapor pressures, ranging within the lower or even mid ppb range at room temperature. So what the sniffer dog's nose detects is the ensemble of the vapors released by <u>all</u> of the components of the explosive package. An odor bouquet is created which acts as a unique fingerprint recognized by the sniffing dog. This "fingerprint" has to be detected by a gas analytical device as well in order to find the explosive. However, as its detectable spectrum of gases is even broader than in case of the dog's nose (e.g. insensitive to CO, methane), the fingerprint might be even more easily detected with an EN.

Tab. 1: Most common explosive agents & their vapor pressures at room temperature [1]

2,4,6-Trinitrotoluene (2,4,6-TNT)	8 ppb
1,3,5-Trinitrohexahydro-1,3,5-triazene (RDX, hexogen)	< 1 ppt
N-Methyl-N,2,4,6-tetranitroaniline (tetryl, CE)	8 ppt
Pentaerythrittetranitrate (pentrite, PETN)	20 ppt
Glycerinnitrate (nitroglycerin, NG)	350 ppb

M. Krausa and A. A. Reznev (eds.),
Vapour and Trace Detection of Explosives for Anti-Terrorism Purposes, 123-131.
© 2004 *Kluwer Academic Publishers. Printed in the Netherlands.*

ANALYTICAL REQUIREMENTS

Conventional high performance gas analytical instruments, such as gas chromatography coupled with mass spec detection are without doubt able to perform a detailed analysis of the vapor released by the explosive package with the required sensitivity. Conventional analysis, however, is discontinuous, takes more than 10min, and its instruments are too large and heavy for mobile onsite use.

The appropriate instrument for onsite search should be capable of online detection, that means repetitive measurement with a response time in the range of only a few seconds or even better. Considering the demand for detecting different explosive agents, as well as a broad variety of volatiles contributing to the characteristic bouquet of gaseous components released by the explosive package, the chosen analytical method should allow the detection of a wide spectrum of gas species. Clearly, the sensitivity required has to be high enough to even detect ppb concentrations. Further, excellent gas discrimination power has to be provided, as the gas emission of an explosive package is always embedded in a variety of gaseous components released by other objects close to it. In addition to the analytical requirements mentioned, such an instrument has to have a low weight, small size and long-term stability to be used in handheld operation. Finally, low power consumption is needed to enable longer periods of mobile operation.

An electronic nose, based on a gas sensor array, provides the basic properties for meeting these requirements: A combination of several gas sensors, with each sensor providing a different sensitivity spectrum, delivers gas characteristic signal patterns of the gases the array is exposed to. These signal patterns enable the distinction between individual gases or gas ensembles. Thus, the fundamental data acquisition and processing of the EN resembles biogenic smell recognition as it detects a complex odor bouquet via an integral perception of the vapor ensemble. If the gas sensors are based on semiconducting oxides (such as SnO_2, WO_3, In_2O_3 and others) which sensitively change their electrical conductivity depending on the composition of the ambient gas phase, detection of nearly all gases is possible, except very inert species such as rare gases or nitrogen [2]. Moreover, a continuous output can be provided with rapid response times. Furthermore, such sensor arrays can be made small and of low weight as well as mechanical robustness is feasible. A micro system design keeps the power consumption low, as metal oxide sensors need to be heated to some 100°C and micro fabrication allows inexpensive production of these sensor arrays.

THE KARLSRUHE MICRONOSE KAMINA

A novel type of EN has been developed at the Institute of Instrumental Analysis of the Karlsruhe Research Center, designed to meet the requirements of consumer products: high gas analytical power has to be combined with inexpensiveness, small dimensions and low power consumption. The development is based on the unique gradient microarrays equipped with a single sputtered thin film of a gas sensitive metal oxide (MOX) which is subdivided by a set of parallel electrode strips, creating 38 gas sensor segments in the standard version (s. Fig. 1) [3].

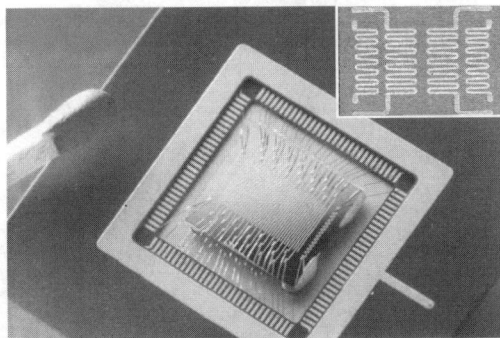

Fig. 1: Gas sensor chip with gradient microarray mounted in its housing. The chip is based on a segmented metal oxide film of which the electrical conductivity is dependent on the ambient gas composition. Two thermoresistor strips beside the metal oxide film serve to control the temperature gradient. Electrical contacts are provided by thermosonic gold wire bonds. The rear side of the chip (upper right) carries 4 meander-shaped heating elements made of platinum to allow inhomogeneous heating of the chip.

The thumbnail-sized gradient microarray chip allows low-cost fabrication as all sensor elements are produced in one step by simply partitioning the sputtered MOX film. A gradient technique differentiates the gas detection selectivity of the individual sensor segments. The thickness of an ultra-thin gas-permeable membrane layer deposited on top of the metal oxide film varies across the array. Additionally, a controlled temperature gradient is maintained across the array. Four heating meanders located on the rear side of the chip are used to operate the micro system in the temperature range of 200-400 °C and –keeping the supplied power to the heaters different- usually a temperature difference of 50 °C is maintained across the array. As both thickness and temperature have a gas dependant influence on the diffusion through the membrane the selectivity of the gas detection gradually change from segment to segment. Moreover, the temperature influence on the gas reaction at the metal oxide / membrane interface also depends on the nature of the gases and hence contributes to the differentiation of the sensor segments.. Therefore the exposure to single gases or gas ensembles (like odors) cause characteristic conductivity patterns to occur at the gradient microarray. The dependence of the conductivity pattern on type and quantity of ambient gases allows gas discrimination and quantification for a limited number of components in a mixture. Hence, this gradient microarray can be applied to realize a sensitive EN system.

A beverage can-sized instrument of only 600 g, the Karlsruhe Micronose (KAMINA), a complete EN system for use in mobile operations has been developed, including with μp-controlled electronics providing operation and on-line data evaluation with 1 Hz (*s. Fig. 2*). The instrument provides high discrimination power for single gases or characteristic gas ensembles and has a response time of a few seconds. For most organic as well as inorganic gases detection limits ≤ 1 ppm are obtained with the standard sputtered MOX films. However, a new generation of microarrays is under development equipped with nanogranular metal oxide layers which seem able to detect even components of a few ppb only. Some first results of microarrays of this type are presented below.

Fig. 2: The KAMINA module shown with the head cover lifted contains the complete electronic nose system including the microarray chip, a small fan for gas sampling and the μp-controlled electronics.

PREPARATION OF NANOGRANULAR METAL OXIDE LAYERS

3" silicon wafers oxidized on both sides were patterned using photolithography to obtain 26 windows, 4 x 8 mm each, for depositing the nanogranular tin dioxide films which served as the basis of the microarray chips. The nanogranular SnO_2 layers were prepared by spin-coating, using an aqueous colloidal dispersion of 15 nm tin dioxide particles mixed with 2-hydroxyethyl cellulose. Because of the poor wettability of the SiO_2 covered wafer surface, it was necessary to add a non-ionic surfactant (dimethylsiloxane-ethylene oxide block copolymer). After deposition, the nanogranular SnO_2-films were dried at 75 °C for 1 h, and then the photoresist was removed with acetone. Subsequently, the SnO_2 films were annealed at 400 °C and 600 °C respectively in ambient atmosphere (*s. Fig. 3*), followed by the deposition of the platinum electrode strips using HF-magnetron sputtering in argon plasma. No difference in the layer properties was found for the two annealing temperatures. Characterization of the surface morphology by Field Emission Scanning Electron Microscopy (FE-SEM) revealed a particle size of about 20 nm for the SnO_2 grains constituting the film after annealing. After dicing the chips were mounted in PGA (Pin Grid Arrays) carriers (*s. Fig.1*) and electrical contacts were made by gold wire bonds.

Fig. 3.: 3" Si wafer with 26 rectangular shaped nanogranular SnO$_2$ detection layers

ANALYTICAL PERFORMANCE TESTS

The gas-sensing properties of microarrays with nanogranular metal oxide layers were investigated by exposures with the test gases isopropanol, benzene, toluene and 2-nitrotoluene, using an automated pulsed gas exposure procedure. The latter alternately exposes the microarray chip to humid air containing defined concentrations of the test gas or to clean air of the same humidity. A few different test gas concentrations are used to obtain the concentration dependence of the microarray's response. In order to investigate the dynamic behavior of the sensor properties abrupt changes of the test gas concentration are realized .

The response to 2-nitrotoluene was especially , because it is known as a by-product of TNT [4]. While the gas concentrations of the other model components were obtained by controlled dilution from certified 100ppm mixtures of the gases in air, 2-nitrotoluene was evaporated from the liquid (boiling point of 225 °C, vapor pressure of approx. 0.1 mbar at 20 °C) using a bubbler operated at 28 °C and subsequently a cooler was applied at 20 °C to accommodate the vapor to room temperature. The 2-nitrotoluene concentrations obtained were checked using a photo-ionization detector (PID) for reference.

RESULTS

The response of the electrical resistances of the 38 sensor segments to a series of test gas pulses containing toluene and clean air in alternation is depicted in *Fig. 4*. As the nanogranular morphology should be investigated concerning its sensing behaviour, the metal oxide layer was tested without coating it with the usual membrane layer (s.o.). The exposure to toluene concentrations led to a decrease of the electrical resistance, which is typical of oxidizable gases. Moreover, in case of the nanogranular SnO$_2$ layer microarray, the differentiation of the 38 sensor segments was brought about by the applied temperature gradient of 250-300 °C. This is seen in *Fig. 4* by the spread of segment resistances.

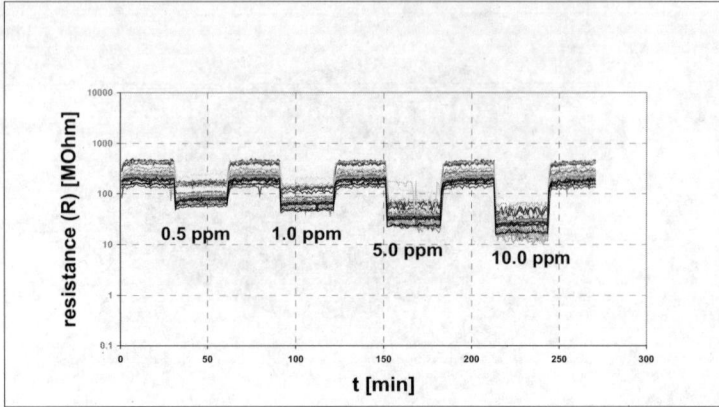

Fig. 4: Resistances of the 38 sensor segments of a SnO2 microarray with nanogranular SnO2 gas detection layer during alternating exposure to toluene contaminated humid air and clean air with the same humidity of 50 %. Toluene gas pulses of 0.5 to 10 ppm were used to test the microarray. The operating temperature range of the microarray was 250-300°C.

The signal of the metal oxide segments S is defined as the relative conductivity change on exposure to the analyte. That is the ratio of the conductivity difference between G(c) of the contaminated air and G(0) determined in clean air with the same humidity divided by G(0).

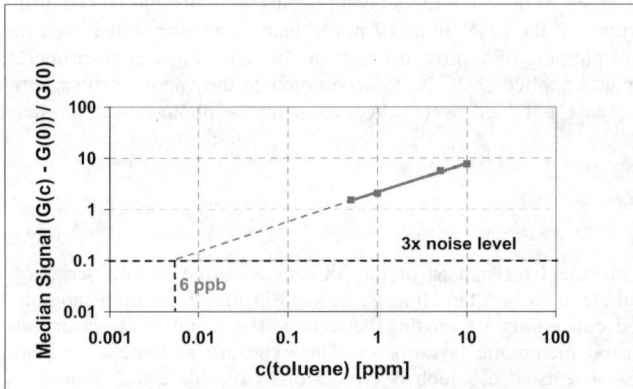

Fig. 5: Calibration curve for toluene using a temperature span of 250-300 °C across the array.

For the sake of convenience, the conductivity rather than the resistance is used, as the relative change in conductivity is conform to the change in concentration for most gases. . The concentration dependence of the segment signal S isshown by the median signal representative of all sensor segments in *Fig. 5* and *Fig.6*.

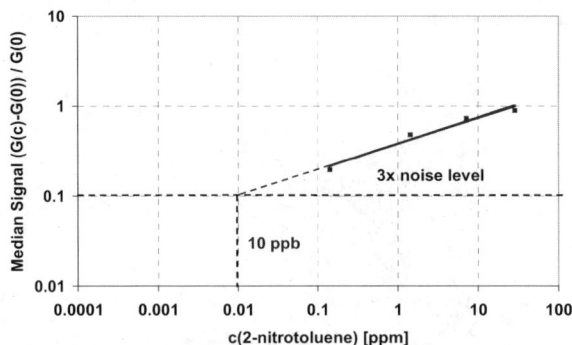

Fig. 6: Calibration curve for 2-nitrotoluene using a temperature span of 250-300 °C across the array

The concentration dependence follows the well-known exponential law $S \sim \alpha C^{\beta}$ (with the gas characteristic constants α and β, $0,5 \bullet \beta \bullet 1$) [2]: Therefore a straight line is obtained in the log-scaled plot of the median signal vs. the concentration. Detection limits are determined by extrapolating the straight line to the significance level $S \bullet 0.1$ defined to be 3 times the noise level. For 2-nitrotoluene, toluene, benzene and 2-propanol detection limits at or below 10 ppb were found (*s. Fig 7*).

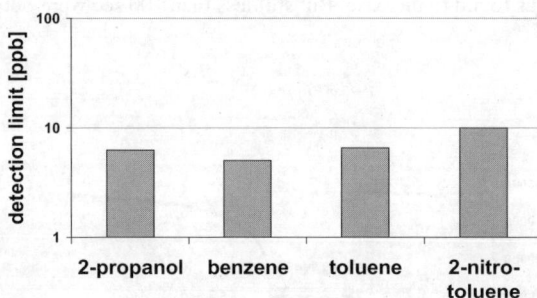

Fig. 7: Detection limits for 2-propanol, benzene, toluene and 2-nitrotoluene exposure using a temperature span of 250-300 °C across the array.

As no membrane coating was used in order to exclusively test the metal oxid layer, the sensor segments could only be differentiated by a temperature gradient. However, even with the applied temperature span of 250-300 °C, the individual sensor segments already all respond in a slightly different manner. The polar plot in *Fig. 8* shows the segment signals as defined before, but normalized to their median for a 10 ppm toluene exposure. Because of the normalization and the signal definition (s.a.), the pattern of the normalized signals for clean air is a circle of unity in this polar plot. Thus, the considerable deviations of the normalized toluene signals from unity show that the temperature gradient can already cause characteristic signal patterns.

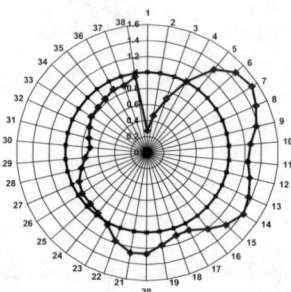

Fig. 8: Normalized conductivity pattern of the 38 sensor segments for 10 ppm toluene exposure using a temperature span of 250-300 °C across the array.

The fast response of the microarray based on nanogranular SnO_2 is shown in *Fig. 9*, depicting the trace of the median resistance obtained for a transient exposure to 2-propanol. Similar to the other tested gases, for a concentration of 100 ppm a response time at which 90% of the final resistance is reached (t_{90}) was below 3 sec. The response time was found concentration dependent. For concentrations below 1 ppm, the response time was found to increase. But still less than 100 sec were obtained.

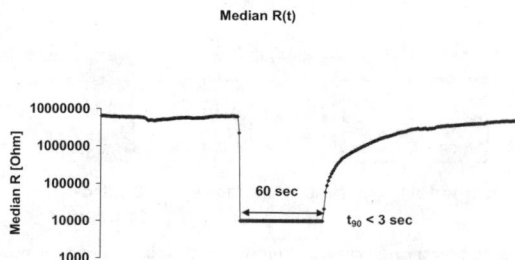

Fig. 9: Response time (t_{90}) for 100 ppm 2-propanol in ambient air using a temperature span of 250-300 °C across the array.

CONCLUSIONS AND TASKS FOR THE FUTURE

The Karlsruhe Micronose KAMINA equipped with a microarray based on a thin layer of 20nm SnO_2 particles has shown a very high gas sensitivity with detection limits in the range of a few ppb up to 10 ppb for aromatic and alcoholic organic compounds. The response times are in the range of some sec up to 100 sec for very low concentrations. Operation with a temperature gradient only, has already shown the ability of the microarray to respond to the gases it is exposed to with gas characteristic signal patterns. These signal patterns form the basis for gas discrimination via the microstructure.

The next step of development will be to deposit a gas permeable SiO_2 layer on top of the metal oxide, carrying a thickness gradient across the array. The gradient membrane will further differentiate the sensor segments and enhance their gas discrimination power.

Although the gas sensitivity might be sufficient to detect the gas release of an explosive package, the identification of such an object in a completely unknown environment that will certainly contain several other sources of gas release remains an enormous challenge. An extremely high gas selectivity is required which could probably only be achieved by an appropriate tuning of all elements of the microarray which contribute to its gas discrimination power: a metal oxide with the appropriate gas selectivity (including dopants for further adaptation), the operating temperature, and the gradients for sensor segment differentiation. However, the resulting optimization of the gas dependent characteristics of the signal patterns should additionally be complemented by improvements in data evaluation to achieve an adequate gas detection system that is well prepared for such a difficult task. Intelligent and fast algorithms are required to attain a reliable deconvolution of the measured signal pattern to extract and identify the signal contribution of the explosive package in the signal ensemble of the total gas composition caused by the combined gas emission of the target object and the items in its vicinity.

However, the complexity of the problem could be considerably reduced when isolating suspicious objects from the environment – just as it is done with hand baggage at airport check points –and separately measured. For instance, if the investigated object to be checked is deposited in a closed box (with a very low evaporation of its own) the requirements for gas discrimination can be met more easily and the gas concentrations resulting from the gas release of the object are also higher due to the confined volume. These investigation scenarios seem to be feasible in much shorter time than the more complicated approach of detecting the explosive package in an arbitrary environment.

REFERENCES

[1] "Technology Against Terrorism: The Federal Effort". OTA-ISC-481. Washington, DC: U.S. Government Printing Office (1991)

[2] www.unibw-muenchen.de/campus/MB/we6/ zeman/explo/zeman_explo_1.htm

[3] Schierbaum, K.D., Göpel, W. (1995)SnO_2 sensors: Current status and future prospects", Sensors & Actuators B **26**, nr. 1-3, 1-12.

[4] Goschnick, J. (2001) An Electronic Nose for Intelligent Consumer Products Based on a Gas Analytical Gradient Microarray; Microelectronic Engineering, **57-58**; 693-703

[5] Köhler, J., Meyer, R. "Explosivstoffe" (in German), Wiley VCH, Weinheim, 1998.

DETECTION OF EXPLOSIVES RESIDUES ON AIRCRAFT BOARDING PASSES

RICHARD SLEEMAN*†, SAMANTHA L. RICHARDS*, I. FLETCHER A. BURTON*,
JOHN G. LUKE*, WILLIAM R. STOTT** AND WILLIAM R. DAVIDSON**

* Mass Spec Analytical Ltd.,Building 20F, Golf Course Lane, Filton, Bristol BS99 7AR,
   United Kingdom
** Sciex, MDS Inc.,71 Four Valley Drive, Concord, Canada, L4K4V8
† Corresponding author (richard.sleeman@msaltd.co.uk)

ABSTRACT

A prototype system for the detection of trace explosives residues on aircraft boarding passes has been developed. The desorption of explosives from the passes was achieved using short wave infrared radiation. The vapours produced were drawn into a triple quadrupole tandem mass spectrometer and were monitored in SRM mode. The infrared unit integrated with the tandem mass spectrometer has demonstrated limits of detection of less than 100 pg for the explosives studied (TNT, NG, PETN and RDX). A background study into the levels of explosives residues on used boarding passes was conducted by analysing over 20,000 boarding passes from a number of airports in England, America and Canada. Traces of explosives were detected on approximately 0.5% of passes analysed. Nitroglycerine contributed to the majority of the positive signals observed. All the signals observed were below 1ng.
An automated unit has been developed after the success of the manual feed system, which incorporated the inclusion of a transport mechanism to move the boarding cards through the desorption window automatically. The desorption efficiency of the system was found to be between 70% and 100%, depending on the thickness of the card used for assessment. The unit was able to handle a throughput rate of 1000 boarding passes per hour and was able to detect between 10pg and 50pg of explosives residue from the surface of the card depending on the compound used.

KEYWORDS

Boarding pass, explosives, desorption, short wave IR, tandem mass spectrometry, atmospheric pressure chemical ionisation, background study, automated transport mechanism.

M. Krausa and A. A. Reznev (eds.),
Vapour and Trace Detection of Explosives for Anti-Terrorism Purposes, 133-142.
© 2004 *Kluwer Academic Publishers. Printed in the Netherlands.*

INTRODUCTION

The technique of tandem mass spectrometry is generally applicable to the trace detection of residues of organic compounds. This paper describes its use in a specific application; that of the detection of explosives residues on aircraft boarding passes. Many of the arguments presented concerning the analytical technique could equally apply to the analysis of carry-on or checked-in baggage, lap-top computers, etc.

The Need for Trace Detection

A requirement exists for personnel screening by means of trace detection methodologies to augment the widespread use of X-ray scanners as these scanners have not provided a complete solution to the terrorist threat. Ideally, one would wish to swab the hands of a passenger to maximise the probability of detection of traces of explosives. That approach is not feasible, however, because it is time consuming and would inconvenience passengers. The analysis of boarding passes has been proposed[1] as one possible alternative for transferring residues from the person to an analytical device. Any technique used for this purpose must be robust and easy to use and, above all, rapid. It should not adversely affect airport operations.

The rationale behind this approach is that those involved with explosives, both directly and indirectly, are likely to become contaminated with detectable amounts of trace residues. These residues are transferred by touch, and therefore, given that passengers must handle their boarding passes, this presents an opportunity for analysis prior to the person boarding an aircraft. For the results to be interpreted in terms of the identification of a potential terrorist threat, it is a prerequisite that few members of the ordinary travelling public be innocently contaminated.

Choice of Analytical Methodologies

When selecting the most appropriate analytical approach to use for the examination of the boarding passes, we considered three steps: the collection, or removal of residues from the pass, the transfer of those residues to the analytical device, and the selection of the most appropriate method of analysis. Our overriding concern was that the approach developed after consideration of each separate step should yield the highest possible overall probability of detection. Nevertheless, we bore in mind the practical considerations of rapidity and robustness.

Dissolution of residues from the passes into a solvent would be highly efficient, but would be time consuming. There would most probably be a need to reduce the volume of solvent prior to analysis. For this reason, this strategy was rejected as a viable sample acquisition approach.

Swabbing was also rejected because of the inefficiency of removal of residues from a paper substrate. The well established approach of thermal desorption was therefore selected as the method of choice. It was also thought that the transfer of explosives to the analyser in the vapour phase would be an efficient process, provided the transfer line was heated and kept as short as possible.

The choice of analytical technique focussed on rapidity, selectivity and sensitivity. We chose tandem mass spectrometry (MS/MS) for this application. Ion Mobility Spectrometry (IMS) was also considered, but rejected on the grounds that the more rapid the technique used, the higher the demands on selectivity. In the airport environment alarm resolution is a major issue, and it is an important consideration that the number of false alarms should be as low as possible. When operating at a high throughput rate of one thousand analyses per hour, a difference of one percent in the number of alarms leads to an additional ten passengers to be investigated further.

MS/MS is also more sensitive than IMS, and this is also an important consideration in boarding pass analysis; data are presented later which illustrate the amounts of material which are likely to be present in real world simulations.

The technique of MS/MS used in Selected Reaction Monitoring (SRM) mode is well established[ii], having first been utilised in this application in the late '70s with the TAGA (Trace Atmospheric Gas Analyser) developed by Sciex which formed the basis of the CONDOR system in the '80s. The deployment of MS/MS devices has, however, suffered from the historical perception that the systems are too complex, expensive and require highly skilled operators. Recent developments have addressed each of these problem areas, and simple-to-use instruments are now available at a reasonable cost. A further benefit, when compared to other forms of commercially available equipment, is that the derived data are forensically acceptable.

Tandem Mass Spectrometry

The process below describes how MS/MS in SRM mode can be used to identify a chemical compound. The analyte enters the source of the mass spectrometer in the presence of dust and debris which together constitute a complex matrix. Within the source the analyte becomes ionised, together with many other species which enter the source in the matrix. By using Atmospheric Pressure Chemical Ionisation (APCI), which is a "soft" ionisation technique, very little fragmentation occurs within the source, and the majority of ions formed from explosives compounds are either M⁻, [M-H]⁻, or simple adducts of these molecular ions. Several other technologies employ APCI, but most use a nickel-63 source to cause ionisation. The use of a radioactive source can lead to practical problems at airports. In the system presented herein, this is replaced by a corona discharge generated at the tip of a needle within the ion source.

The mass spectrometer has two mass filters. The first allows the selection of ions with the appropriate mass to charge ratios. All other ions have unstable trajectories within the quadrupole field and are lost at this stage. Only the selected ions of interest, together with isobaric ions (those having the same mass to charge ratio which could potentially interfere and produce false positive alarms) proceed to the next stage. These ions are caused to collide with molecules of nitrogen gas, which is bled into the collision region at low pressure. The impact causes collisionally induced dissociation (CID) to occur, generating a number of "product ions" in a reproducible manner. The second mass filter is tuned to allow the passage only of certain product ions which are selected as being representative of the compound of interest. If the choice of product ions is appropriate, ions generated from the decomposition of isobaric species can be effectively eliminated and hence false positive alarms can be ruled out. Compounds are thus identified by selected gas phase ion reactions from a known "precursor ion" to a number (usually two is sufficient) of specified "product ions". The whole process takes less than 10ms and the mass spectrometer is set to cycle repeatedly between the analysis of a range of compounds (currently six in routine operation although up to sixteen has been demonstrated). Thus the system operates in quasi real time.

This approach has a number of benefits, the obvious ones being extremely high specificity (which in practice means a low false positive alarm rate), high sensitivity and rapidity. Further benefits include the fact that the background can be continuously monitored to ascertain the chemical noise floor, and that the approach is readily adaptable to new threats, such as the emergence of new terrorist explosives. In most cases, SRM transitions for a new compound can be selected within a few hours of investigation to give a high probability of detection.

EXPERIMENTAL AND RESULTS

Fig. 1 depicts an in-house built short wave infrared thermal desorber system for the analysis of boarding passes coupled to a Sciex API365 triple quadrupole tandem mass spectrometer. It is important to note that the boarding passes are heated directly and there is no sample pre-treatment of any kind.

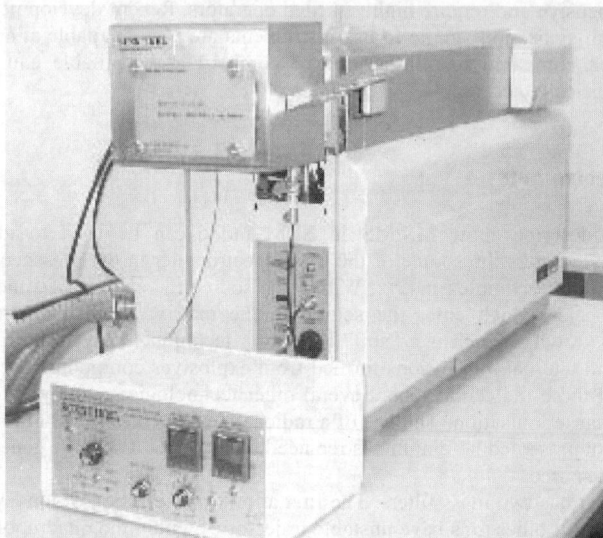

Fig. 1: Boarding Pass Analyser and Tandem Mass Spectrometer

Tab. 1 illustrates the gas phase ion transitions used for the detection of four common explosive threats, NG, PETN, RDX and TNT. Two transitions are selected to be representative of each compound. Note that in the case of TNT, two characteristic transitions from the M⁻ ion can be used. With, NG, RDX and PETN, however, few specific transitions are observed in their respective product ion spectra. For these compounds, extra certainty in identification can be provided by the use of chloride adducts[iii]. In order to produce these adduct ions, dichloromethane vapour is bled into the source. Adducts are formed by the addition of either $^{35}Cl^-$ or $^{37}Cl^-$ in a 3:1 ratio according to the relative abundances of the two stable isotopes of chlorine; ^{35}Cl and ^{37}Cl. Not only does this provide more potential transitions to monitor, but the transitions from these adduct species to any particular product ion, say m/z 46 (NO_2^-), also occur in that same characteristic ratio.

Tab. 1: SRM Transitions for Explosive Detection

Compound	Precursor ion (m/z)	⇨	Product ion (m/z)
NG	264 $(M + ^{37}Cl)^-$	⇨	46 (NO_2^-)
	262 $(M + ^{35}Cl)^-$	⇨	46 (NO_2^-)
PETN	353 $(M + ^{37}Cl)^-$	⇨	46 (NO_2^-)
	351 $(M + ^{35}Cl)^-$	⇨	46 (NO_2^-)
RDX	259 $(M + ^{37}Cl)^-$	⇨	46 (NO_2^-)
	257 $(M + ^{35}Cl)^-$	⇨	46 (NO_2^-)
TNT	227 (M^-)	⇨	210
	227 (M^-)	⇨	197

Fig. 2 illustrates the detection of 100pg NG from a boarding pass. The material was introduced in solution in methanol via syringe. Although this is a good method of introducing a known amount of material, a certain amount of "wicking" occurs, where the solution soaks into the fibres of the pass, and small crystals form. This is somewhat different from real world situations where the explosives would be deposited on the surface of the pass as larger crystals in a matrix of finger grease. When thermal desorption is used, the detection probability is likely to be higher for larger crystals on the surface than it will be for smaller crystals deep within the paper substrate. Thus this represents a difficult detection scenario.

Note the (approximate) 3:1 ratio of the species containing the stable isotopes of chlorine.

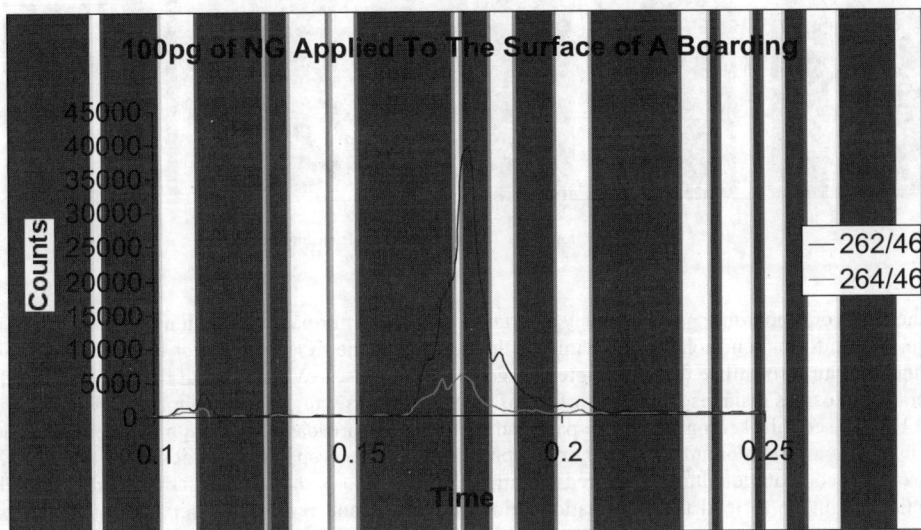

Fig. 2: Detection of NG on a Boarding Pass

Fig. 3 depicts the analysis of a boarding pass to which 500pg of both RDX and PETN had been applied via syringe. Both compounds were simultaneously detected and in each case a 3:1 ratio was observed between the two ion pairs which incorporated a chloride adduct.

138

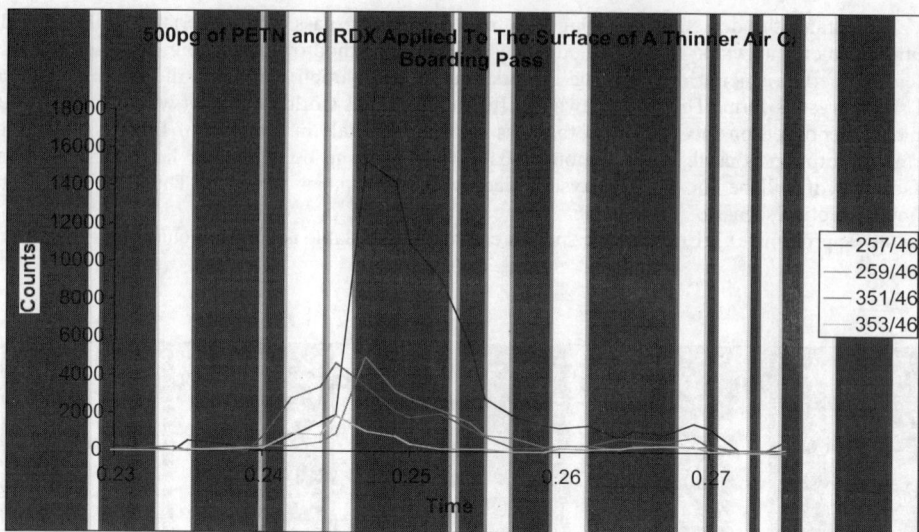

Fig. 3: Detection of PETN and RDX on a Boarding Pass

In the real world boarding passes convey a certain amount of information in both magnetic and printed form. This information includes the name of the passenger, the destination, seat numbers, etc. Some airlines and airports utilise thermal printers to generate the passes. When thermal desorption is used to liberate explosives residues from the surface of a pass, the pass will absorb radiant energy and the ink will be released, blackening the whole pass and rendering it unreadable. Although this neither renders the magnetic information unreadable, nor compromises detection capability, it does mean that this type of pass is incompatible with the infrared radiation thermal desorption approach advocated. A possible solution might be to read the information prior to analysis and reprint another pass after analysis, although it may prove cheaper simply to utilise other forms of printing when generating the passes. A further alternative might be to introduce some additional form of security pass separate to the boarding pass itself.

Detection efficiency was influenced by the thickness of the card used. This was estimated by repeatedly desorbing spiked cards. *Fig .4* illustrates the sequential desorption of a pass spiked with NG. On the first insertion, at approximately 0.13 minutes or 7.8s, a large response was recorded. This was followed (at approximately 0.23 minutes, or 13.8s) by a period in which residual explosive in the machine was purged. The same pass was reinserted at 0.30 minutes (18s) and a further peak was observed. No peak was observed from a third insertion at 0.45 minutes (27s). Measurement of the respective peak areas revealed that approximately 70% of the material detected was removed from the pass and transported to the analyser on the first analysis. This rose to ca. 100% when thinner card was used.

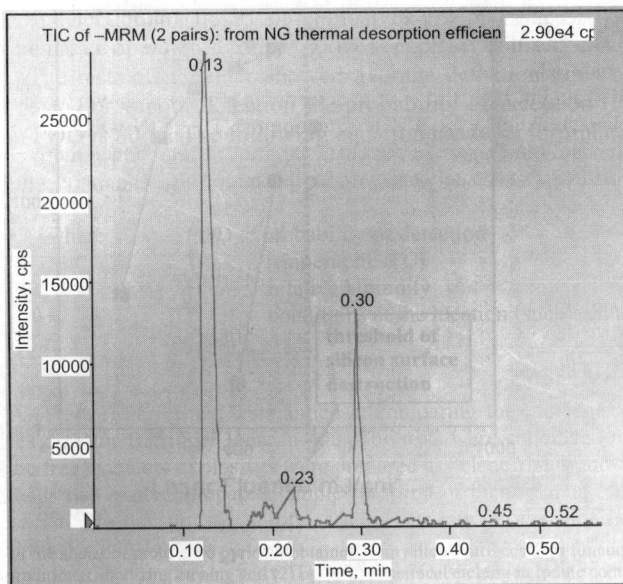

Fig. 4: Repeated Desorptions of a Pass Spiked with NG

Similar results were obtained with PETN and TNT, where thick card gave 40% efficiency for both explosives, rising to 70% for thinner card. Results for RDX were 30% for thick card and 60% for thinner card.

Other features to notice in this figure are the desorption peak width, which corresponds to a period of less than 2s, and the time between analyses of approximately 10s, even when purging took place. Note also that virtually no "memory effect" can be observed.

Various approaches have been applied to modelling the likely amounts of material to be found in the event of a terrorist threat. These have included the application of explosives slurries to surfaces, etc. Unfortunately these approaches are not applicable to all explosives types and are not fully representative; finger grease and sweat are generally not present. Neither are they applicable to all surface types; hard plastics, fabrics, etc.

In this study, a pragmatic approach was adopted where real explosives were used. Explosives were handled directly by a "bomb-maker", who would form an improvised device. These devices were subsequently handled by an "accomplice", who would take this device and place it in an item of baggage. A "bystander" (or "dupe") would then handle the baggage. Boarding passes handled by persons from each of these three categories would then represent "easy", "moderate" and "difficult" detection scenarios respectively.

In the first attempts, semtex was used to simulate a "moderate" detection scenario. The "bomb-maker" handled semtex and then washed his hands thoroughly, using soap, hot water and paper towels. A clean glass vial was then handled by this person for approximately thirty seconds. A second person, the "accomplice", then handled the vial, and subsequently touched a boarding pass. Both RDX and PETN were successfully detected when this boarding pass was analysed.

This detection is illustrated in *Fig. 5* where the detection of PETN is shown by the transitions in green and orange in the characteristic 3:1 ratio. The detection of RDX is shown (*Fig. 5b*, scale enlarged) by the transitions in blue and pink, again in the characteristic 3:1 ratio. Once more note that the rise and fall of the peak occurs in approximately 6s, even for this large "hit", and there is little memory effect.

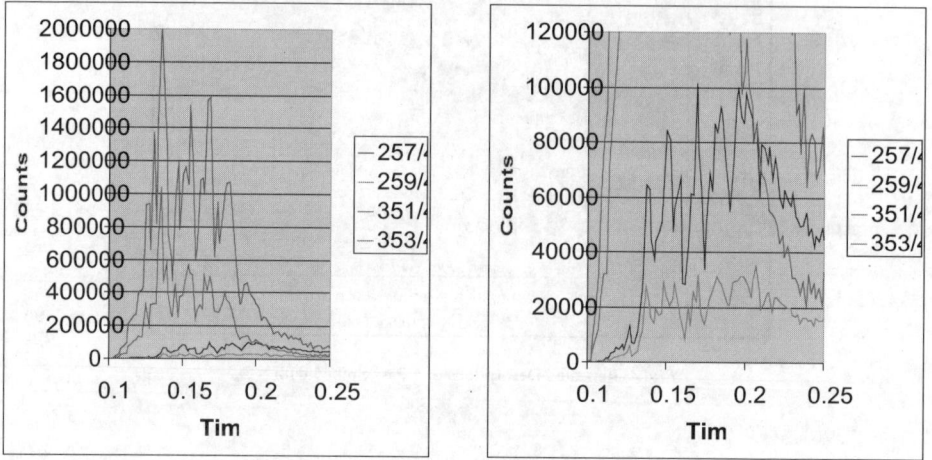

Fig. 5: Semtex Bomb-making Simulation
(a) (Left) All Transitions Shown (b) (Right) Expanded Scale

Fig. 6 shows the analytical device, complete with all peripherals.

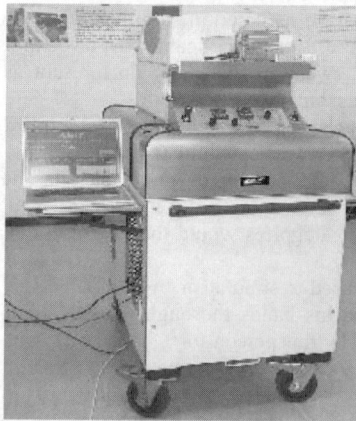

Fig. 6: The Analytical Device

Interpretation of Findings

No security measure can be effective unless the results obtained can be put into perspective. In the case of trace detection of explosives residues, there is a possibility that innocent people may be found to be contaminated through the use of explosives or explosives compounds for purposes unconnected with terrorism. Such uses may include the use of explosives in the mining or demolition industries, military use, or the use of explosives compounds in medical applications or for recreational purposes by those with certain sexual proclivities. Contamination of the skin with explosives derived via any of these mechanisms is indistinguishable (by this method) from that caused by terrorist involvement. Therefore the probability of other types of explosive use contributing to a "hit" must be assessed before any system can be brought into operation in security screening. This information will help predict the likely number of "nuisance alarms". Note that these are not "false alarms", because the instrument has responded correctly to the presence of explosive; the fact is, however, that a detection of the presence of explosives does not necessarily correlate with a threat.

Some twenty thousand boarding passes were collected from the UK, USA and Canada. These were analysed to determine the throughput rate, the robustness of the analytical system and the proportion of passes contaminated with various compounds.

Fig. 7 illustrates the proportion of "hits" for each of the four explosive types sought. Each of the explosives was detected to some extent. The majority of these hits were for NG, which may be explained by its use as a heart drug. The overall "hit" rate was less than half of one percent. Each of the detections were estimated to be below 1 ng of material.

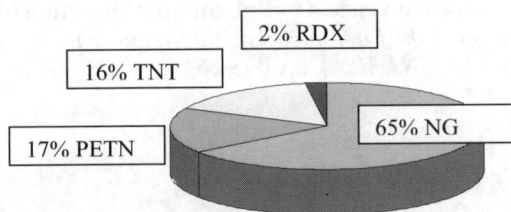

Fig. 7: Proportion of Detections by Compound

CONCLUSIONS

In conclusion, bulk detection strategies do not yet provide a comprehensive solution to the terrorist threat. The analysis of boarding passes provides a convenient way of detecting terrorist threats. Tandem mass spectrometry allows analysis to take place at a rate which does not inconvenience passengers or airport operation. A number of explosives may be detected concomitantly with an acceptable nuisance alarm rate and very low false alarm rates.

ACKNOWLEDGEMENTS

Transport Canada are thanked for financial support.

REFERENCES

[i] Richards, S.L.; Sleeman, R.; Burton, I.F.A.; Luke, J.G.; Carter, G.T.; Stott, W.R. and Davidson, W.R. (2001). **The Detection of Explosives Residues from Boarding Passes.** In: *"Proceedings of the Seventh International Symposium on the Analysis and Detection of Explosives" (Edinburgh, Scotland),* pp. 60-65.

[ii] Davidson, W.R; Stott, W.R.; Akery, A.K. and Sleeman, R. (1991). **The Role of Mass Spectrometry in the Detection of Explosives.** In: *"Proceedings of the First International Symposium on Explosives Detection Technology" (Atlantic City, NJ, USA),* pp. 663-671.

[iii] Davidson, W.R; Thomson, B.A.; Sakuma, T.; Stott, W.R.; Akery, A.K. and Sleeman, R. (1991). **Modifications to the Ionization Process to Enhance the Detection of Explosives by API/MS/MS.** In: *"Proceedings of the First International Symposium on Explosives Detection Technology" (Atlantic City, NJ, USA),* pp. 653-662.

FREIGHT SCREENING – TRACE DETECTION

BILL HALKETT

The FLORENCE project, Security Processes Ltd., 64 Chorley Road, Bispham, Ormskirk, Lancs., L 40 3 SL, UK

KEYWORDS / ABSTRACT:

trace detection / freight screening / trucks / air freight / threats / particle detection / vapour detection / high volatility / low volatility / airflow / low pressure zone / RASCO / turbulent flow / FLORENCE project / aerodynamic modelling

Freight screening using trace detection is difficult but sample acquisition is the key to any trace detection operation: no sample = no detection.
This paper proposes a solution to the understanding of sample acquisition from freight which uses experience gained from dogs trained to the UK RASCO standard and its application to technology based detectors. This raises the issue of how auditable and repeatable calibration and training materials can be produced for these detection / analysis systems.

ACRONYMS

Acronym	Definition
EGDN	Ethylglycoldinitrate
HARC	Houston Advanced Research Center
NASA	National Aeronautics and Space Administration
PASS	Pressure Activated Sampling System (CyTerra Corp product)
Pd	Probability of Detection
PFA	Probability of False Alarm
RASCO	Remote Acquisition of Samples for Canine Olfaction – a method of training dogs using gram or sub-gram amounts of explosive.
R&D	Research & Development
RDX	Cyclotrimethylenetrinitramine (a military high explosive)
TCD	Trace Chemical Detection
WMD	Weapon(s) of Mass Destruction

M. Krausa and A. A. Reznev (eds.),
Vapour and Trace Detection of Explosives for Anti-Terrorism Purposes, 143-152.
© 2004 *Kluwer Academic Publishers. Printed in the Netherlands.*

INTRODUCTION

This paper sets out to describe the results of an investigation into the use of trace chemical detection systems for screening freight. The imaging technologies, such as x and gamma ray systems, provide users with data which can often be interpreted in several ways and the only to resolve a possible alarm is to hand search the load. Trace chemical detection (TCD) can, if used correctly, provide corroborative data which will allow users to reduce the number of false alarms and consequently interruptions to the flow of commerce.

BACKGROUND

Freight vehicles and containers are used to transport all kinds of contraband and a failure to find one type probably means a failure to find all threats. Typically these might be:

- Explosives,
- Drugs,
- Weapons,
- Weapons of Mass Destruction (WMD), and
- People.

All of these threats can be carried in freight and if one can be smuggled through, then so can the others. In other words all contraband could be a potential terrorist threat.

It must not be forgotten that airfreight is the big black hole in aviation security and that 60% to 70% of all air cargo is carried on passenger aircraft. It is important to find a reliable way to screen airfreight without splitting it apart into the small items that x-ray operators can interpret.

For the purposes of this paper, radiological materials will be ignored as there are remote sensing methods for these. The other WMD's present other detection challenges and trace detection may not be the optimal technology with which to find them: a leaky canister of a chemical or biological agents will leave recognisable trails behind them – the difference being that one will be more immediately identifiable than the other.

As a matter of fact, freight screening then becomes an investigation into the acquisition of samples for trace chemical analysis from freight vehicles and containers because without samples, trace chemical analysis systems will not produce a result because if there is no sample there can be no detection, regardless of system sensitivity:

No sample = no detection

The problem is that very little work is being, or has been, done to maximise the material available for analysis. It's very unglamorous work which may appear to be a low-tech challenge compared to the development of new detector technologies but the fact is that the workable sample acquisition solutions need high tech input from engineers and scientists. Trying to get R&D funds for sample acquisition studies is very difficult and it is very difficult for commercial concerns of the size of the trace detection companies to fund this work out of revenues – let alone get access to threat materials in representative quantities for adequate data collection and process verification.

THE CHALLENGE

Explosives have a wide range of vapour pressures from about 250,000,000 parts per trillion (ppt) for EGDN to 6 ppt for RDX and this is a large part of the challenge which faces extraction of samples from large enclosed spaces. Chemical warfare agents fall into the high volatility range. Biological warfare agents do not have volatility as such and so the detection and identification process is more complex and currently worryingly time consuming.

As a consequence of the wide range of volatilities, sample acquisition processes vary from "sniffing" to wiping. Sniffing is a problem for the low volatility materials because of the difficulty in obtaining enough vapour for analysis and wiping is difficult for high volatility explosives because of the poor retention characteristics of the common collection materials.

Typical sampling devices range from vacuum cleaner type suction devices to manual wiping – See *Fig.1* for some examples used for conventional sampling activities. Each of these images shows a particle sampling process. They are all manual and very local; i.e., the sampling process is close to or in touch with the object being sampled. This is not very useful for freight. These samplers depend upon a particle being available for collection at the place where the wiping or suction device is being used. For the sample acquisition process to be successful, these particles have to be transported from the source to a place from where it can be collected. That's not to say that it will be collected but at least it may be available. The source may be a long way from the collection point or packaged in some way.

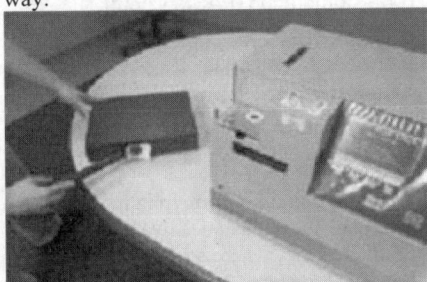

Smiths Detection (Ionscan 400B system)

GE Iontrack (Itemiser system)

Biosensor Applications (BIOSENS system)

Thermo Electron Corporation (Egis system)

Fig. 1: Typical manual sampling devices

There has been some development directed at automatic sample acquisition by RAY Buechler (Discovery) and CyTerra Corporation (PASS). Both of these systems use variations on variation of pressure and / or temperature and in the case of the Discovery system, vibration. *Fig. 2* shows these systems.

RAY Buechler (Discovery system)   CyTerra Corporation (PASSsystem)

Fig. 2: Automatic sampling systems

However, none of this helps to resolve the question of obtaining samples from freight vehicles or containers. From time to time US Customs, Revenue Canada and the Houston Advanced Research Center (HARC) prepare a test site at Fort Huachuca in the Southern USA where several containers are loaded with bulk cocaine. Manufacturers and developers of TCD systems are invited to test their sample acquisition and analysis equipment. Typical of the sampling devices deployed are those shown in *Fig. 3* from SecureTec and Biosensor Applications. In many ways the Ft Huachuca tests are benign in the sense that the ambient temperature is high (>25°C typically) and the level of clutter around the blocks of cocaine is not very complex. Positive detections were made using these installations but it's a big step from cocaine and most other narcotics to explosives when one considers the physical structure of most explosives let alone the availability of vapour and / or particles in colder conditions and in complex loads.

SecureTec   Biosensor Applications

Fig. 3: Ft Huachuca tests

RASCO

Remote Acquisition of Samples for Canine Olfaction (RASCO) was developed in the late 1990's by the UK Government for protection of possible infrastructure targets. It has long been known that dogs have extraordinary scent based detection capability but typical operational dogs are active, get tired and have to be rotated regularly. RASCO set out to use passive dogs to process samples acquired by another process and it is the success of RASCO that fundamentally affects our perception and understanding of sample acquisition from freight loads.

Fig. 4 shows a schematic of the RASCO process. A probe is inserted into the load bay of a truck or container; air is extracted through a filter capsule; the filter is then, with several blanks, presented to a RASCO trained dog which will indicate if it detects a programmed substance.

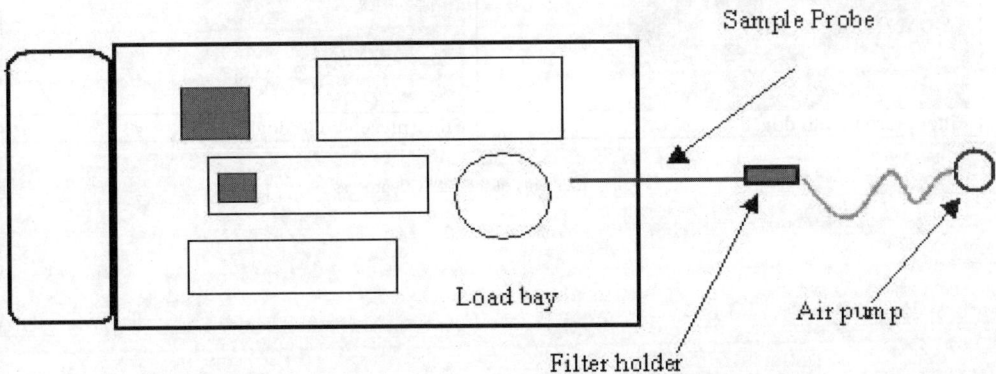

Fig. 4: RASCO sampling schematic

The probe is about 2m long with an 8 mm inside diameter. Typically, the pump runs for between 3 and 8 minutes at 90 to 100 litres / minute; i.e. a maximum of 800 litres. The filter is plastic netting wrapped around a former and it is all assembled into a capsule. It is worth noting that the filter is not pushed into the load area – it is connected in series between the rigid probe and flexible hose which in turn is connected to the air pump. This fact is a problem for traditional TCD thinking when dealing with low volatility explosives because these explosives are known to be sticky and will adhere to any surface if at all possible. Using simple Reynolds number calculations, it is clear that the air flow in the probe is laminar – ie the walls are not scoured by turbulent airflow but as RASCO probes are regularly re-used without apparently causing false alarms, this in turn indicates that the substances which trigger a RASCO dog are not left behind in the tube where they might cause a false negative or if released at a later time, cause a false alarm.

There are two qualified RASCO companies in the UK - Kent K9 Ltd and Chilport UK Ltd and, although they have different training and sample presentation methods, the end result is the same – high performance. Because we know that RASCO works, it is clear that one or more detectable substances exist and are available for collection. The images at *Fig. 5* show elements of the RASCO process.

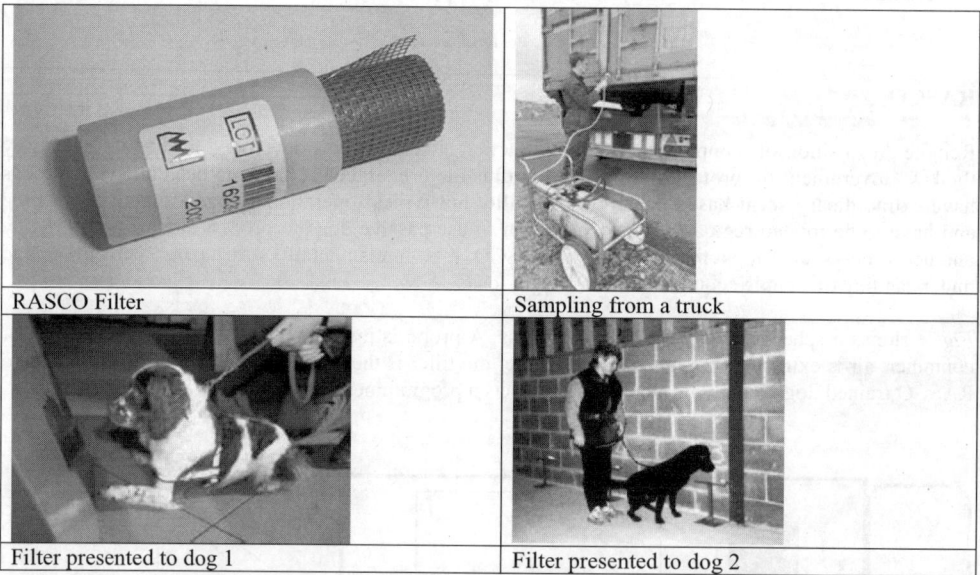

RASCO Filter	Sampling from a truck
Filter presented to dog 1	Filter presented to dog 2

Fig. 5: Elements of the RASCO process

Traditional TCD sample acquisition thinking would indicate that:
(i) There should not be very much material to capture: possibly fractions of a pico gram per litre at high ambient temperatures. In any case 800 litres of air from a cargo space of up to 37,000 litres is not much and, since the flow will come from the volume with least resistance to airflow, most of it will come from the volume closest to the rear of the load space and even the leaking seals of the doors.
(ii) The probe tube should be internally contaminated after a positive detection because of the "sticky" nature of these substances.
(iii) Particle detection is the most probable method of sample acquisition from low volatility explosives but in RASCO, the filter has a wide mesh and even taking account of overlapping of the strands, the effective pore size will be measured in large fractions of a millimetre rather than microns. In other words many particles will flow straight through the filter.

RASCO has been tested extensively in the UK under widely varying weather conditions and on the basis of more than 2000 analyses, has shown Probability of Detection (Pd) of > 98% with false alarms at less than 1%. These are impressive figures but why and how does RASCO work and can those answers be used to improve sample acquisition for electronic TCD's?

SAMPLING FROM FREIGHT

So how does RASCO work? The RASCO sampling process clearly obtains enough material for a trained dog to detect – but where does the sample come from and why is it that a short probe – perhaps only 1m in a total freight bay length of 13m or more – obtains enough material of the right kind to allow a dog to make a high reliability detection decision? Answers to these questions should indicate the way forward for all freight screening using TCD.

The images in *Fig. 6* which show trucks in motion have one factor in common: the curtain side is being sucked in. It is clearly an aerodynamic effect – so what does aerodynamic modelling show? Work carried out for fuel economy studies show that air over the trailer and at the sides and back of the truck is turbulent – See *Fig. 7a* and *7b*. It is clear from these figures that turbulence at the rear of the truck feeds air into the wake.

Fig. 6: Trucks in Motion

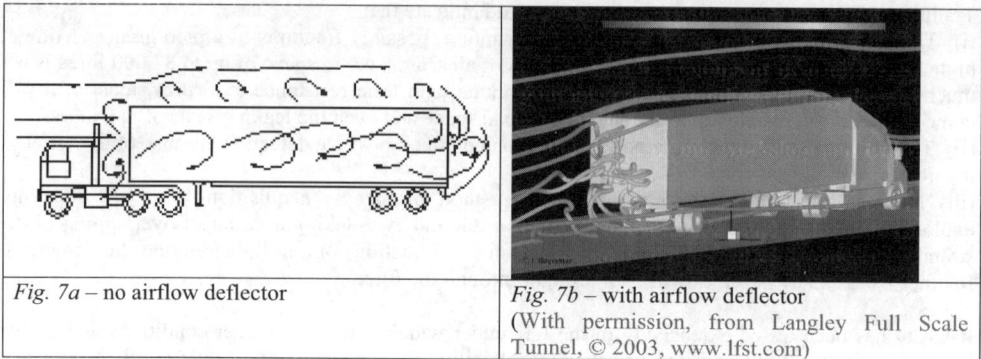

| Fig. 7a – no airflow deflector | Fig. 7b – with airflow deflector (With permission, from Langley Full Scale Tunnel, © 2003, www.lfst.com) |

Fig. 7: Truck aerodynamics

MECHANISM

It is proposed that the mechanism which generates samples at the rear of trucks for RASCO sampling and analysis is purely aerodynamic. The low pressure zone at the rear of the container or truck extracts air from within the load space. Air is moved inside the load space from all points within that space and if there are particles or vapours available at some point in the load space they will be transported to the low pressure zone. Some of these vapours and particles will flow out through the door seals to be taken away in the wake or to adhere to the outside of the doors and some will adhere to the surfaces at the rear of the load space – the insides of the doors, floor and walls.

This solution does not answer the question about whether the dogs are detecting the actual low volatility materials or other higher volatility components which are <u>always</u> present but this will discussed later.

There will be no aerodynamic difference between curtain sided and aluminium or steel sided trucks and containers. They will all have low pressure zones which will cause enough leakage of air through the load bay to pull dust and particles and even – who knows – vapours to the back of the truck. Because this is a continuous process, a bulk source is required to sustain the sample availability; i.e. traces left behind from previous loads of explosives or other contraband. Laser spectrographic analysis, carried out by the UK Ministry of Defence, of deposits on the rear of trucks confirms that traces are present.

This analysis raises the question: why dogs and not technology? It is generally recognised that dogs have high sensitivity and selectivity. Work carried out by a number of research establishments indicates that dogs use a bouquet of scents to identify a target material. They certainly have much higher analysis capability than all of the current and developmental electronic TCD's and from an operational standpoint, it does not matter if dogs are actually detecting RDX, say, or another compound which just happens to always be present when RDX is present.

What is interesting from the total TCD viewpoint is the RASCO sample acquisition process. Having resolved the question of why volatile and non-volatile materials will be transported to the rear of the load bay by internal air movement, the next question is: why does the RASCO process work reliably for low volatility explosives. These compounds should adsorb readily to most surfaces and freight

loads are complex mechanically and chemically. The stickiness of these compounds should also mean that they adsorb to the inside of the probe tube causing random false negatives from retained material and random false positives from released material but this does not happen and at least one of the two approved RASCO suppliers re-uses probes after hits without any adverse affect on performance.
Does this mean that dogs detect higher volatility components? If so, what are they?
These are the questions which prompted the FLORENCE project.

The FLORENCE Project The FLORENCE Project seeks to:

- Explore in more detail the sample transfer and acquisition mechanisms which allow RASCO dogs to perform at these very high levels
- Establish processes and protocols which will produce reliable and reproducible calibration standards.
- Establish audit standards for RASCO users.
- Establish whether RASCO can be adapted to electronic detection systems by establishing the bouquet of volatile substances being used by dogs to identify explosives.

In this work the project workers are not attempting to solve the question of what a dog detects and why the dogs work better on some days than others, but if RASCO based technology is to be used for freight screening, then it is necessary to know how well it is working at any one time. As an example, recent trials work with different air freight configurations seems to show that, although detection rates were still high, false alarms increased. These results could be due to a number of factors but it is probable that the major effects are:

- The much smaller quantity of explosive, and hence surface area, of explosive;
- The test pallets and containers have not been transported to the test site and so explosive residues have not been transported within the load.

Consequently the dogs are trying too hard for the reward.

CONCLUSIONS

- Sample acquisition is a fundamental requirement for effective trace detection.
- Sample acquisition is not well understood or developed.
- RASCO is effective for truck screening but not well understood.
- RASCO type processes require quality control and audit trail processes.
- Technology based detectors will be better if they detect what the RASCO dogs detect.
- Aerodynamic modelling of threat objects will optimise sample collection.

Finally, it must not be forgotten that threat detection has great similarities to manufacturing processes. It involves materials, people, processing machines and control systems and procedures. In this case the materials may be explosives; the people will be the bomb makers or law enforcement agencies; the processing machines will be detectors of some type and the control systems will be the procedures which make the whole system work. The bomb maker cannot work without explosives or initiators and the detectors will be trace or bulk. The detectors need data. In the case of the x-ray systems they are the signals from the detector array, in the case of trace detection, it is the trace itself. In the case of x-rays systems the sample acquisition is automatic but in most cases the data processing is manual – the operator's brain. In the case of trace, the sample is normally acquired manually and the analysis is automatic. In each case the reliability of detection is only as good as the weak link and that is usually the manual part. Until we understand each element of the threat detection process, we will not be able to achieve the high operational detection rates needed to protect society.

ACKNOWLEDGEMENTS

Thanks for information and comments are due to the following companies and organisations:

Lt Col (Ret[d]) M P Groves, UK
Biosensor Applications AB, Sweden
Chilport (UK) Ltd, UK
CyTerra Corporation, USA
GE Ion Track, USA
Kent K9 Ltd, UK
RAY Buechler, Israel
SecureTec, Germany
Smiths Detection, USA
Thermo Electron Corporation, USA
NASA Full Scale Tunnel Facility, Langley, USA
US Customs Service, USA